Understanding Evolution

Why do the debates about evolution persist, despite the plentiful evidence for it? Breaking down the notion that public resistance to evolution is strictly due to its perceived conflict with religion, this concise book shows that evolution is in fact a counterintuitive idea that is difficult to understand. Kostas Kampourakis, an experienced science educator, takes an insightful, interdisciplinary approach, providing an introduction to evolutionary theory written with clarity and thoughtful reasoning. Topics discussed include evolution in the public sphere, evolution and religion, the conceptual obstacles to understanding evolution, the development of Darwin's theory, the most important evolutionary concepts, as well as evolution and the nature of science. *Understanding Evolution* presents evolutionary theory with a lucidity and vision that readers will quickly appreciate, and is intended for anyone wanting an accessible and concise guide to evolution.

Kostas Kampourakis is the author and editor of books about evolution, genetics, philosophy, and history of science, and the editor of the Cambridge University Press book series *Understanding Life*. He is a former editor-in-chief of the journal *Science & Education*, and the book series *Science: Philosophy, History and Education*. He is currently a researcher at the University of Geneva, where he also teaches at the Section of Biology and the University Institute for Teacher Education (http://kampourakis.com).

The ***Understanding Life*** **Series** is for anyone wanting an engaging and concise way into a key biological topic. Offering a multi-disciplinary perspective, these accessible guides address common misconceptions and misunderstandings in a thoughtful way to help stimulate debate and encourage a more in-depth understanding. Written by leading thinkers in each field, these books are for anyone wanting an expert overview that will enable a deeper understanding of each topic.

Series Editor: Kostas Kampourakis http://kampourakis.com/

Forthcoming titles:

Understanding Evolution	Kostas Kampourakis	9781108746083
Understanding Coronavirus	Raul Rabadan	9781108826716
Understanding Evo-Devo	Wallace Arthur	9781108819466
Understanding Genes	Kostas Kampourakis	9781108812825
Understanding Development	Alessandro Minelli	9781108799232
Understanding DNA Ancestry	Sheldon Krimsky	9781108816038
Understanding Creationism	Glenn Branch	9781108927505

Understanding Evolution

KOSTAS KAMPOURAKIS
University of Geneva

CAMBRIDGE
UNIVERSITY PRESS

University Printing House, Cambridge CB2 8BS, United Kingdom

One Liberty Plaza, 20th Floor, New York, NY 10006, USA

477 Williamstown Road, Port Melbourne, VIC 3207, Australia

314–321, 3rd Floor, Plot 3, Splendor Forum, Jasola District Centre,
New Delhi – 110025, India

79 Anson Road, #06–04/06, Singapore 079906

Cambridge University Press is part of the University of Cambridge.

It furthers the University's mission by disseminating knowledge in the pursuit of education, learning, and research at the highest international levels of excellence.

www.cambridge.org
Information on this title: www.cambridge.org/9781108478694
DOI: 10.1017/9781108778565

First published 2020

Printed in the United Kingdom by TJ International Ltd, Padstow Cornwall

A catalogue record for this publication is available from the British Library.

ISBN 978-1-108-47869-4 Hardback
ISBN 978-1-108-74608-3 Paperback

"While other books explain what is wrong with the popular attacks on evolution – e.g. creationism, or Intelligent Design – this concise book addresses the fundamental question: *why* do people fail to accept evolution? This is like going deep to the causes of the illness, while others just try to lower the fever.

Kampourakis argues convincingly that teleology, rather than theology, is the most important obstacle to understanding evolution. It is not just matter of science vs. religion.

This welcome book is a long overdue argument about the cultural and psychological roots of the widespread misunderstandings of evolution. It opposes scientism – the claim that evolution, or science in general, can bring an end to our questions, worries, and concerns; and, at the same time, it argues that evolutionary theory does not deprive our life of meaning."

Alessandro Minelli, University of Padova, Italy, and author of *Plant Evolutionary Developmental Biology*

"A well-known philosopher of biology once wrote that evolutionary theory seems so simple that almost anyone can misunderstand it. In this heartfelt yet thoughtful book, Kostas Kampourakis essentially turns that sentiment on its head. The author's words on philosophy and science may well lead readers to conclude that, although evolution can be counterintuitive and complex, almost anyone can understand it, with suitable reason and evidence. Kampourakis' treatment should be especially enlightening for those who are wrestling with the acceptance of evolution as truth."

John C. Avise, Distinguished Professor of Ecology and Evolution, University of California–Irvine, and author of *Evolutionary Pathways in Nature: A Phylogenetic Approach*.

"*Understanding Evolution* by Kostas Kampourakis deserves a wide readership. It is a sensitive introduction to evolutionary theory itself, as well as its public image and its philosophical implications. It shows the very great importance of the father of the subject, Charles Darwin; sets the disputes with religion in context; and suggests that the evidence is overwhelming but that no reader need feel threatened. It is fair and comprehensive, lively without being heavy-handed, and judicious in its judgments. Read it yourself, and get a copy for your family and your friends!"

Michael Ruse, Florida State University, and editor of *The Cambridge Encyclopedia of Darwin and Evolutionary Thought*

" ... there are plenty of good books on evolution ... So why another one? Because, argues Kampourakis, *contra* a widespread assumption among educators, the biological theory of evolution is actually counterintuitive, and, if not properly taught, it immediately runs into incomprehension and generates conceptual confusion.... Scientific theories are dynamic, ever changing, perpetually incomplete and open to revision ... The more the public at large understands this, the better off we will be, and books like Kampourakis' certainly make a valuable contribution to nudging us in that desirable direction."

Massimo Pigliucci, K. D. Irani Professor of Philosophy, The City College of New York

"In *Understanding Evolution*, Kostas Kampourakis provides not only a masterly exposition of the elements of evolution but also a compelling explanation of why the topic is so difficult to understand. Informed by up-to-date biology as well as by state-of-the-art historical, philosophical, and psychological scholarship, the book is a concise and considered treatment that deserves the attention of anybody interested in evolution.

Glenn Branch, Deputy Director, National Center for Science Education

"This is, without a doubt, the best book available that deals with what is often referred to as a straightforward dichotomy of 'science versus religion.' Not so is the central message of this outstanding, well-written and organized treatment of the nature of evidence, of theory – especially evolutionary theory – of the ramifications of evolution throughout society, and, perhaps most importantly, of why it is so difficult for so many to accept the evidence for evolution. Aimed at a general audience, this book should be read by all who struggle with the logic and consequences of the theory of evolution."

Brian K. Hall, FRSC, University Research Professor Emeritus, Dalhousie University, Canada

"'How extremely stupid not to have thought of that!', Huxley exclaimed, when he first learnt of evolution by natural selection. But Darwin's great insight is not at all obvious – in many ways, it is rather counterintuitive. The odds of experiencing that 'Aha!' moment are vastly improved by teachers like Kostas Kampourakis. Always attentive to conceptual obstacles, Kampourakis helps the reader to grasp the core of evolutionary theory – from evo-devo to genetic drift – to show just how rich and exciting it is."

Tobias Uller, Professor of Evolutionary Biology, Lund University, Sweden

"In *Understanding Evolution*, the science educator Kostas Kampourakis offers an accessible but sophisticated analysis of the topic, combining historical, scientific, philosophical, and psychological factors. Though not ignoring religious concerns, especially those related to design, he focuses instead on the numerous difficulties associated with understanding evolution. Specialists and dilettantes alike will learn much from this volume."

Ronald L. Numbers, Hilldale Professor Emeritus of the History of Science and Medicine, University of Wisconsin–Madison

"*Understanding Evolution* is an outstanding resource for students, teachers, scientists, and journalists. It sets an impressive new standard for the field by integrating current findings from biology, psychology, and the philosophy of science. Using clear and compelling examples, Kampourakis uncovers the roots of our intuitions about the living world, and shatters widespread myths about why resistance to evolutionary ideas is prevalent. Readers will be rewarded with new tools for fostering scientific literacy, and fresh insights into one of the most profound biological ideas."

Ross H. Nehm, Professor of Ecology and Evolution, Stony Brook University, and Editor-in-Chief, *Evolution: Education and Outreach*

"This volume addresses an important and timely issue – why does the concept of evolution encounter such resistance? – and provides a clear, original, and richly informative answer. Taking an interdisciplinary approach, the author reveals persistent conceptual obstacles that have broad implications for the nature of scientific understanding in the world today."

Susan A. Gelman, Heinz Werner Distinguished University Professor of Psychology and Linguistics, University of Michigan, USA

To my wife, Katerina, and our children, Mirka and Giorgos, for turning an inherently purposeless life into a deeply meaningful one.

Contents

Foreword

Back in 2014, Cambridge University Press published Kostas Kampourakis' original book *Understanding Evolution*. He wrote it as a textbook, aiming to bridge the gap between the concepts and conceptual obstacles to understanding evolution. The response was overwhelmingly positive, with enthusiastic endorsements from philosophers and historians of science, to biologists and science educators.

When Kostas and I came to discuss a potential new edition of his book, we agreed that it was important to ensure it was as widely accessible as possible. We discussed how we could achieve this and what the barriers to understanding were. From this emerged the idea of re-writing the book more fundamentally so that it would serve students, but also a broader, general audience. We agreed that the driving force of the book would be to identify and unpick the conceptual obstacles to understanding. From here arose our thinking of the potential value in applying this to a wide range of topics across the life sciences. And so the *Understanding Life* Series came to be.

Our vision for the series is to provide concise, accessible guides to key topics, written by leading thinkers in the field and focusing on the common misconceptions and misunderstandings that are potential barriers to gaining a deeper understanding.

The response from potential authors to this series concept has been wonderfully positive, as you will see from the list of forthcoming titles. We look forward to working with these authors and many more in the future, to bring you this series of exceptional titles. It is a joy to work with

Kostas Kampourakis on this series – his energy, ideas, insights and ability to tease out the barriers to understanding and learning on any given topic know no bounds.

Dr Katrina Halliday
Executive Publisher, Life Sciences
Cambridge University Press

Preface: There is More to Resistance to Evolution than Religion

Evolutionary theory is the central theory of biology. It explains the unity of life by documenting how extant and extinct species share a common ancestry. It also explains the diversity of life by describing how species have evolved from ancestral ones through natural processes (a "species" can be defined as a group of individuals that can interbreed and produce fertile offspring, although this definition overlooks the complexities of microbial life). Today, an evolutionary perspective is dominant in many of the most active fields of biological research and also provides important insights in medical, agricultural, and conservation studies and applications. The evidence for evolution is vast and comes from several different disciplines, such as paleontology, systematics, developmental biology, and genomics, which makes scientists consider evolution to be a fact of life. All in all, evolutionary theory is a powerful scientific theory that organizes and provides coherence to our understanding of life. As Theodosius Dobzhansky, an important evolutionary geneticist of the twentieth century, famously stated, without evolution biology seems like a pile of sundry facts that make no meaningful picture as a whole.

Yet the idea of evolution has been, and continues to be, enormously debated in the public sphere. Various polls around the world have shown that there is a rather low public acceptance of evolutionary theory (discussed in Chapter 1), in many cases due to its perceived conflict with religious beliefs and worldviews (discussed in Chapter 2). Related to this is the relatively high acceptance of creationist ideas. In general, creationism is the belief that God created the universe, including the Earth and humans, through a series of miracles. Young-Earth creationists perceive the world to have been created in six days of 24 hours each, some time within the last 10 000 years, whereas

Old-Earth creationists accept the scientific account of the age of the Earth but still believe that the creation of life took place through a series of miraculous interventions. A recent version of creationism is intelligent design (ID), the proponents of which consider, for instance, the vertebrate eye or the bacterial flagellum as irreducibly complex systems: they become non-functional if a part is removed. Therefore, they cannot have gradually evolved through evolution by natural selection, because any form lacking a part would be non-functional and would die out. Therefore, the argument goes, such systems can only have been created for their current roles by an intelligent agent, and so they stand as evidence for ID. As these arguments have been debunked repeatedly, I do not discuss them in the present book.

Many excellent books on evolution have been written, including sound arguments and suggestive evidence that shows not only that evolution is a fact of life, but also that evolutionary theory provides the best scientific explanation for all biological phenomena. However, the authors of most of these books seem to take for granted that it is simple for their readers to understand evolution. Therefore, it seems to be assumed that all people need are books that present arguments and evidence for evolution and/or against creationism. But if such books exist, why then do the public debates about evolution persist? Why is it the case that many people reject evolution or question its validity, despite the evidence for it and its enormous explanatory power in contemporary biological research?

In my view, there is a gap in the existing literature on this topic. Evolution is a rather counterintuitive idea (from a psychological point of view), and it should not be taken for granted that it is easy for all, or even most, people to understand it. There is ample research in psychology that supports the conclusion that resistance to scientific theories may be due to intuitions that generate preconceptions about the natural world, which in turn make scientific findings seem unnatural and counterintuitive. Such intuitions are never completely overwritten, despite even expert scientific knowledge. As a result, the preconceptions that people hold make evolutionary concepts difficult to understand. An additional problem is that people may misinterpret the implications of evolutionary theory for their lives, and may also extend these to questions beyond the realm of science. What is necessary is that people realize that evolutionary theory, like all scientific theories, is a means to understand the natural world, and nothing more. It is also a theory that can be put to the test and not something to which we should dogmatically subscribe.

I have therefore written this book in an attempt to fill this gap in the literature, while also trying to present evolutionary theory in a comprehensible manner. To achieve this, I rely not only on evolutionary biology, but also on conceptual development research and on scholarship from both the history and the philosophy of biology. My main intention is to clearly describe the core concepts of evolutionary theory (in Chapters 5 and 6). However, before attempting this, I am being explicit about the obstacles that affect understanding of evolution (in Chapter 3), suggesting that the low percentage of acceptance of evolution is in part due to a lack of the required understanding. I also show that even Darwin himself had to undergo a process of conceptual change (Chapter 4). Thus, this book explains both what evolution is, and why it is difficult to understand. Given that evolution is a rather counter-intuitive idea, whether people understand evolution or not *is* a major issue, and one that may have been overlooked in the debates surrounding evolution. Throughout the book I also address some common misunderstandings about evolution, which are also summarized at the end of the book.

I should note at this point that I do not overlook the cultural, religious, worldview, and other issues implicated in the problem of the public acceptance of evolution (the term "public" is used vaguely in the present book to refer to all ordinary people). I am aware that there are powerful social factors at work, especially among fundamentalist religious believers, which may have nothing to do with conceptual issues. These people usually associate evolutionary theory with a set of liberal values that they perceive as a threat to their own conservative values. They also usually perceive evolutionary theory as a threat to important social and moral issues (see Chapter 7). However, research in the history of science and in sociology has shown that the relation between science and religion has been, and continues to be, a complex one rather than a simple dichotomy. But as many excellent treatments of the interplay between science and cultural, social, religious, and worldview factors have already been written, I have decided to rather focus on conceptual issues. Due to these, there is more to resistance to evolutionary theory than religious belief.

A note of caution: In Chapters 2 and 3 I present the findings of various studies on children's and adults' design teleology and psychological essentialism conceptions. While reading these sections, you should keep in mind two important limitations of those studies: (1) these are short-scale studies with small sample sizes; and (2) they have involved children mostly from the USA. As a result, generalizations are not easy to make, but I consider their findings as important to report. The main reason for this is that all these studies together support the conclusion that the main conceptual obstacle to understanding evolution is the "design stance": the tendency to perceive "design" in nature and elsewhere. Even though many of us might take as self-evident what design is about, defining it is far from simple. In this book, I consider design as the property of a whole to have parts organized in such a harmonious manner so as to efficiently perform a particular function. By this definition, whatever exhibits design should also reflect the intentions of its designer: The arrangement of parts in such a way that makes a function possible should automatically reflect the intentions of the designer related to the performance of that function. In this view, all biological characteristics, that is, all recognizable features of an organism (which can exist in a variety of states and at several levels from the molecular to the organismal), are also perceived to be the outcome of intentional design.

The metaphor of design has been a popular one in evolutionary biology, and scholars have argued that Darwin explained how there can be design without a designer. Simply put, according to this view, natural selection can bring about the design that we perceive in the structure of organisms, as those organisms that exhibit the best "designs" are also those that are better at surviving and reproducing, and therefore those that pass on their characteristics to the next generation. However, personifying natural selection by thinking of it as a blind watchmaker, as Richard Dawkins has suggested, or even as a tinkerer, as François Jacob once suggested, can be misleading. Dawkins and Jacob were certainly aware, and explicit, that these are just metaphors. But metaphors can be misunderstood because people may pay attention to the part of a metaphor that makes more sense to them, and overlook its limitations. Therefore, stating that biologists can study the structure of organisms *as if* that structure exhibits some kind of design (which we would expect to see in the work of an engineer or a tinkerer), might make people pay attention to this design and its implications about the existence of a function or of a designer. But as I explain in Chapter 3, even though only artifacts exhibit design because they are intentionally created for a purpose, we tend to perceive organisms (especially their parts) in the same way as the parts of artifacts. This is why the metaphor of design in biology had better be avoided.

The main aim of this book is to help readers understand evolution. But because evolution is a counterintuitive idea, this can only happen after readers realize why evolution is difficult to understand. I hope that after reading this book, you will not only realize which obstacles make evolution difficult to understand, but will also be guided to overcome these obstacles yourselves.

Acknowledgments

There are always many people an author can thank for their intellectual contributions and their support in writing a book. But there is no one else that deserves to be acknowledged more in this case than Katrina Halliday, executive publisher for the life sciences at Cambridge University Press. Neither this book as you see it, nor the book series to which it belongs, would have existed without the insight and support of Katrina. The first edition of the present book, published in 2014, was very well received and was commended for its unique contribution and its readability (see excerpts from and links to the reviews at http://kampourakis.com/understanding-evolution). Yet, that was still an academic book. Thanks to Katrina, we now have this revised and updated, but also concise, version that I hope you will appreciate.

I am indebted to many scholars for their ideas, comments, and suggestions: John Avise, Francisco Ayala, Glenn Branch, John Hedley Brooke, David Depew, Patrick Forber, Jim Lennox , Alan Love, Kevin McCain, Sandro Minelli, Robert Nola, Ron Numbers, Greg Radick, Henk de Regt, Karl Rosengren, Michael Ruse, Mike Shank, Elliott Sober, Paul Thagard, John Wilkins, and Tobias Uller. Finally, I am grateful to Ross Nehm and Michael Reiss for useful discussions on topics related to understanding evolution over the years.

My interest in understanding evolution goes back in time. I am indebted to Vasso Zogza, my PhD advisor, who helped me understand that conceptual development research has a lot to contribute to understanding science concepts. I am also indebted to my old friend Giorgos Malamis, who guided me through my first forays into the vast literature of philosophy and history of science when I was an undergraduate student. Finally, I am grateful to Eleftherios Geitonas, founder and director of Geitonas School, and to all my former colleagues there who supported my research during the 12 years that I worked there as a biology teacher.

I am also grateful to Olivia Boult and Sam Fearnley at Cambridge University Press for their work toward the publication of this book, as well as to Gary Smith for his meticulous copy-editing. Finally, I thank Mihalis Makropoulos and Sinos Giokas, who notified me about some minor issues that they identified while working on the Greek translation of the 2014 edition.

Over the years I have extensively discussed many of the issues raised in the book with my wife, Katerina, my best friend and companion in life, who also has a background in the life sciences. Her thoughts, comments, and fierce criticism have always been valuable. Moreover, while writing I was thinking that this book should be appropriate for our children, Mirka and Giorgos, to read when they grow up. Existential questions will come up at some point and I wanted to be able to give them this book in order to read about how scientists study the natural world and what they can, and cannot, conclude about it. Thus, I have written this book with my own children and their intellectual/conceptual development in mind.

For being a source of inspiration and for making me feel sentimentally rich, I dedicate this book to my family: my wife and our children for turning an inherently purposeless life into a deeply meaningful one.

1 The Public Acceptance of Evolution

Evolution in the Polls

What is evolution? The term might refer either to the *fact* that species have changed over the course of eons, or to the *process* by which this change has taken place, resulting in their exquisite adaptations and their outstandingly common features. All organisms are related to one another because they have descended from a common ancestor through natural processes that have produced new life forms from preexisting ones. It is important to note that evolution has been taking place on Earth for billions of years. Consequently, although it is still taking place now, much of the information about it comes from the past. Evolutionary scientists do not have a direct view of the past, but they can infer past events from what they currently observe. Overall, there is ample evidence for evolution in fossils, anatomy, biogeography, and DNA.

However, the idea of evolution in general and of human evolution in particular is usually misrepresented in the public sphere, with illustrations such as the one in Figure 1.1. There are two main problems with this representation of human evolution. First, it portrays evolution as a linear process in which each one of the species changes into another one. However, evolution is more accurately represented as a branching process, not a linear one. Second, this representation shows humans evolving from apes that exist today. This is misleading too, because a species cannot evolve from other contemporary species. What is actually happening is that humans and apes share common ancestors, from which they have evolved independently, like branches starting from a common shoot. But before explaining evolution in detail, it is interesting to consider its public image.

Figure 1.1 One of the usual misrepresentations of human evolution as a series of transitions among coexisting species.

The public acceptance of evolution has been the focus of various polls. Polls are a useful means to acquire a snapshot of what people think about various issues; some are conducted at the national level, whereas others are international. In the latter case, it is possible to compare attitudes and knowledge of people living in different countries, under the condition that the samples studied are representative of the respective populations. Organizations such as Eurobarometer, Gallup, Pew, Ipsos, and others are supposed to provide valid and reliable data on what people think about various topics. There are many interesting conclusions one can draw from such polls; however, this should be done with caution. There are at least three kinds of issues that one must keep in mind when considering the results of these polls. These are: (1) methodological; (2) conceptual; and (3) inferential.

Methodological issues have to do with whether the research questionnaires used actually measure what they are supposed to measure (validity), and with whether this is done in a reliable manner (reliability). To give a simple example, if I use my ruler to measure a length of 10.5 cm, I need to know if what I measure is indeed 10.5 cm (validity), and if I obtain this very same measurement every time I use this ruler (reliability). I write this chapter under the assumption that there are no such issues in the reports of Eurobarometer, Gallup, Pew, and Ipsos that I consider. This entails that I take for granted that the questionnaires used in the respective studies were correctly understood by the participants, who thus provided responses about the topics they were expected to think about and who would provide the same response on different occasions. However, it is possible that survey questions have been constructed in ways that potentially lead to biases and distortions of the actual views held by those surveyed. This can happen, for example, due to a focus on human evolution, which might make respondents feel uncomfortable – someone who chose a religiously justified answer might be concerned that they would be considered ignorant due to the lack of an opportunity to defend this choice and to present oneself as knowledgeable in this matter.

Conceptual issues have to do with the content of the questions; more specifically they relate to whether the concepts used are accurately defined, and to whether the questions cover all the relevant conceptual variation. For instance, in the "UK BBC Horizon: A War On Science" poll, participants were asked which of the following three statements best described their view of the origin and development of life:

- The "evolution theory" says that humankind has developed over millions of years from less advanced forms of life. God had no part in this process.
- The "creationism theory" says that God created humankind pretty much in his/her present form at one time within the last 10,000 years.
- The "intelligent design" theory says that certain features of living things are best explained by the intervention of a supernatural being, e.g., God.

As it has been correctly pointed out, there is no choice that might refer to the views described as theistic evolution (evolution guided by God) and deistic evolution (evolution initiated by God without any further intervention). This entails that religious participants might have been forced to choose either the creationism or the intelligent design option, even though these options might not accurately reflect their own thinking. In this sense, this study might yield a higher number of creationists than there actually are.

Finally, what I have called inferential issues have to do with the inferences that one can or cannot make, and do or do not make, from the poll data. Whereas looking at participants' responses to individual questions is often used as the basis for conclusions, I argue that one should rather look at participants' responses to different questions of the same study, as well to questions of different studies, in order to make better-grounded inferences as to what participants think. For instance, a common conclusion from polls is that in highly religious countries the acceptance of evolution is lower than it is in more secular countries. Thus, one might be tempted to infer that the more religious a country is, the less accepted evolution will be. However, when one looks into the details, there is not a simple evolution/religion dichotomy, and what emerges is a more complicated picture. In this chapter, I focus mostly on conceptual and inferential issues, leaving the methodological issues aside, because I am interested in the conceptual content of the questions and in how the emerging results might be (mis)interpreted.

Some articles presenting results of evolution-focused polls around the world have attracted considerable attention. For instance, a 2006 article published in the prestigious journal *Science* compared attitudes in various countries to the statement that "Human beings, as we know them, developed from earlier species of animals." Participants were asked whether the statement was true or false, whether they were not sure or did not know. It was found that about 25 percent of participants from Turkey and about 40 percent of participants from the USA considered the foregoing statement as true, whereas this was the case for more than 80 percent of participants from Iceland, Denmark, Sweden, and France. Another article, published a couple of years later – again in *Science* – reported on the findings of a study in predominantly Muslim countries, asking participants the following question: "Do you agree or disagree with Darwin's theory of evolution?" Not many people agreed that Darwin's theory is probably or almost certainly true: 16 percent in Indonesia, 14 percent in Pakistan, 8 percent in Egypt, 11 percent in Malaysia, 22 percent in Turkey, and 37 percent in Kazakhstan. Such findings seem to show a clear pattern: People in more religious countries are less likely to accept evolution than people in more secular countries, as well as that people in predominantly Christian countries are more likely to accept evolution than people in predominantly Muslim countries. However, if one looks at the details of these polls, there is more than that, as I show in the subsequent sections.

Evolution Polls in Europe

During January–February 2005, data from 32 countries were collected, through personal interviews, by the European Commission. The findings were published in the Eurobarometer survey 63.1 in June 2005 (this is where much of the data for the 2006 *Science* article previously discussed came from).

The study involved participants from the 25 (at that time) member states of the European Union, as well as from Bulgaria, Romania, Croatia, Turkey, Iceland, Norway, and Switzerland. Two reports were released. The first one was titled *Special Eurobarometer 224: Europeans, Science and Technology*, and the other was titled *Special Eurobarometer 225: Social Values, Science and Technology*. One of the questions asked in the survey concerned the statement: "Human beings, as we know them, developed from earlier species of animals." Participants were given the choices "true," "false," or "don't know." The findings are presented in Figure 1.2.

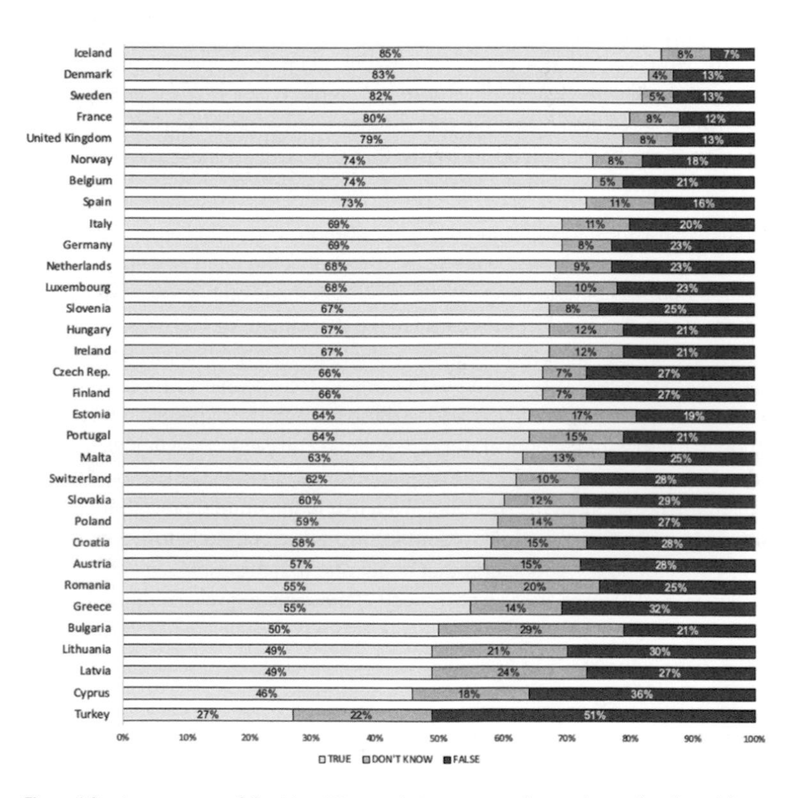

Figure 1.2 Acceptance of the idea "Human beings, as we know them, developed from earlier species of animals" in European countries and Turkey.

The first noteworthy issue is the content of the survey statement itself. Strictly speaking, the statement is incorrect because no species can "develop" from an earlier species. The term "development" is currently used in the life sciences to refer to individual life cycles and within-generation time spans. It is rather "evolution" that refers to populations and time spans across generations. Therefore, the statement should instead have been written as "Human beings, as we know them, *evolved* from earlier species of animals." It is unclear whether replacing the verb "evolve" with the verb "develop" was done accidentally, or intentionally in order to refrain from using an evolution-related word. One might indeed argue that if using an e-word is a sensitive issue, one had better refrain from using it and replace it with less sensitive words. However, such a choice raises important conceptual issues. If you think about this, the word "development" implies a more goal-directed process than "evolution." Stating that humans have developed from earlier species might be perceived to imply that this was an inevitable outcome; however, human evolution was far from inevitable.

Conceptual issues notwithstanding, what else do we see in Figure 1.2? There are multiple ways to look at the results. One is that the majority of participants in all European countries accepts the idea of humans originating from animal predecessors, an idea rejected by half of the participants in Turkey. This sounds like good news for Europe. However, if you look closely at the results, you will also see that between one in five and one in four people in most European countries reject this idea. If you add to these the number of people who do not know what to think, overall about one in three Europeans does not accept the idea of human origins from animal predecessors. One might still be pleased with these results though, especially given that in the same survey about one in three participants in the 25 EU countries agreed with the statement that "The Sun goes around the Earth" and that about one in five people agreed with the statement that "The earliest humans lived at the same time as the dinosaurs." In other words, there are fundamental issues related to science literacy that do not have to do with the idea of evolution only. Some people may just be ignorant about science in general, and not antievolutionists.

Nevertheless, a usual concern whenever there are people who seem not to accept the idea of evolution is that their religious worldviews may be responsible for this. Another question asked in the survey was the following: "Which of these statements comes closest to your beliefs?" Participants could choose among the following statements: "I believe there is a God"; "I believe there is some sort of spirit or life force"; "I don't believe there is any sort of spirit, God or life force"; "I don't know." As is evident in Figure 1.3, there is variation in the belief in the existence of God in the various countries. However, some kind of spirituality is also quite widespread, and as a result less than one in three participants in all countries expressed their disbelief in the existence of God or some spiritual entity.

A question then comes up naturally: Is there a connection between the belief in the existence of God and the low acceptance of evolution? Figure 1.4 presents together the results already presented in Figures 1.2 and 1.3 about the number of people who believe in the existence of God and the number of people who considered the statement that "Human beings, as we know them, developed from earlier species of animals" as being false.

Two important inferences can be made from Figure 1.4. The first one is that not all people who believe in the existence of God also consider the idea of humans originating from animal predecessors as false. What is even more interesting, though, is that, with the exception of Turkey, the number of participants rejecting the idea of humans originating from animal predecessors is 20–30 percent in most countries, both in the more "religious" and in the less "religious" ones. The results were quite different in Turkey, which is also the only predominantly Muslim country. These findings support the conclusion that Christianity, which is the major religion in Europe, does not necessarily relate to opposition to the idea of evolution. However, the findings from polls in the USA provide a very different picture.

Evolution Polls in the USA

For a period of 37 years, between 1982 and 2019, Gallup has been conducting surveys in the USA, asking participants the following question: "Which of the following statements comes closest to your views on the origin and development of human beings?" Participants could choose one among the following options:

- Human beings have developed over millions of years from less advanced forms of life, but God guided this process.
- Human beings have developed over millions of years from less advanced forms of life, but God had no part in this process.

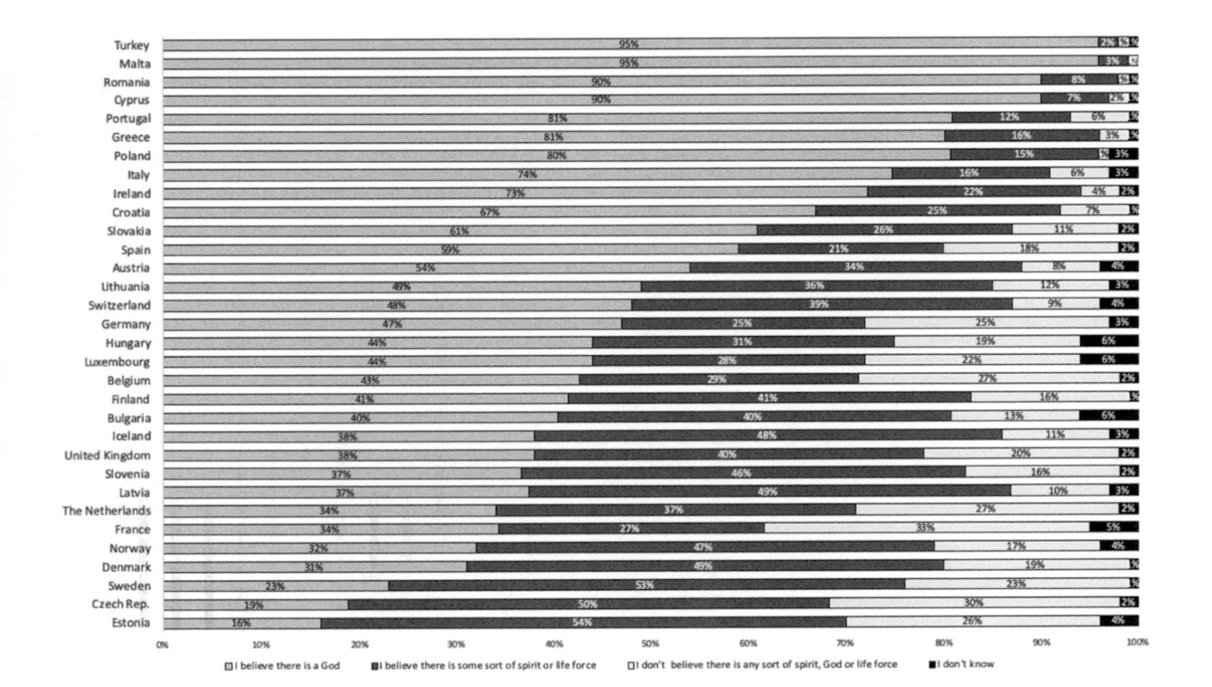

Figure 1.3 Belief in the existence of God or some sort of spirit or life force in European countries and Turkey.

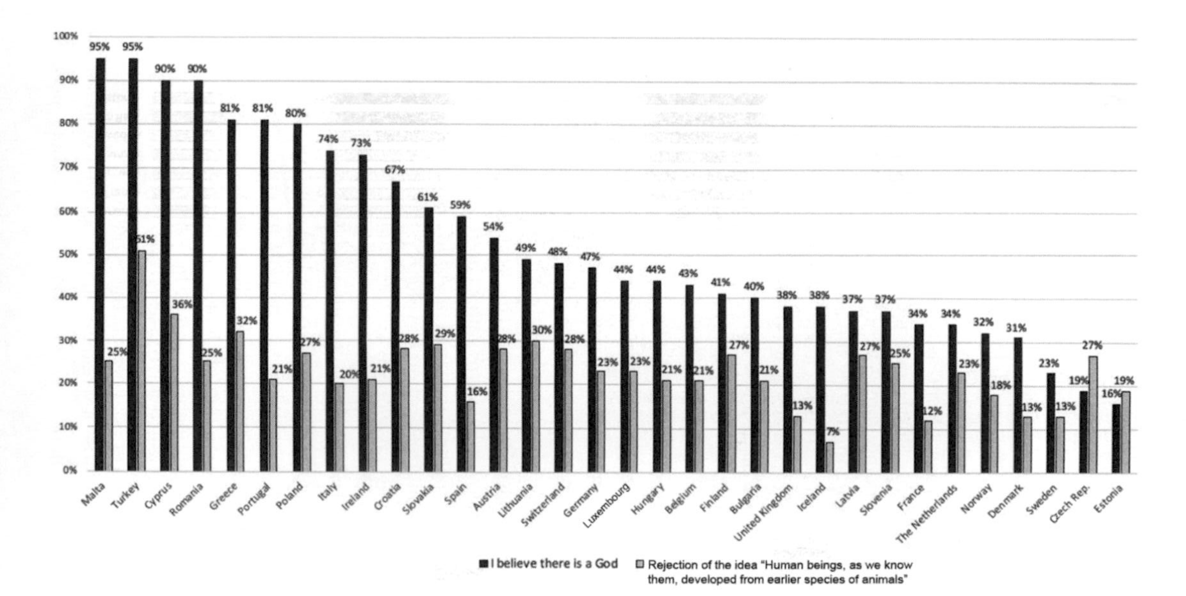

Figure 1.4 Rejection of the idea "Human beings, as we know them, developed from earlier species of animals" and belief in the existence of God in European countries and Turkey.

- God created human beings pretty much in their present form at one time within the last 10 000 years or so.

Before discussing the results of this survey, it is worth considering for a moment the conceptual content of this question in order to better understand what it is really asking for. The first point to note is that the word "evolution" does not appear in this question. Rather, participants are asked about the "origin" and (as in the Eurobarometer survey) the "development" of human beings. As already explained, the word "development" is not the appropriate one in this context, but "origin" definitely is. Simply put, then, the question asked participants where human beings come from. One of the options was that God created humans in their present form during the last 10,000 years. Let us call this the "creationism" explanation. This is clearly an explanation that is not in agreement with the scientific findings that our divergence from our last common ancestor with apes took place a few million years ago. The other two options accept the idea of evolution over millions of years from other forms of life; however, they differ in whether God was involved in this process or not. In one of them, we are told that it was God who guided the process, and this can be described as "theistic evolution." In the other case, we are told that God had no part in this process. This phrase leaves open whether or not God set the conditions for evolution to occur, an idea often described as "deistic evolution." But as the statement is not explicit about this, and – most importantly – as the researchers themselves consider it as "the 'secular' viewpoint, meaning that humans evolved from lower life forms without any divine intervention," we can refer to this as the "evolution" explanation. All in all, participants thus had to choose among a natural explanation ("evolution") and two supernatural ones ("theistic evolution" and "creationism").

What are the results? For 37 years, "creationism" has been the most popular explanation for the origin of humans in the USA, always being selected by more than 40 percent of participants. The only exception was 2017, when 38 percent of participants chose this explanation and another 38 percent chose the "theistic evolution" option. It must be noted that 2017 was the first time that these two options were chosen by the same number of participants. What must also be noted is that 2019 was the year when "evolution" reached the highest percentage ever, 22 percent. This is not much, but it is a lot better than the 9 percent that chose it when this survey began. Could this indicate a trend toward less acceptance of "creationism" and more acceptance of "evolution"? Perhaps, but this remains to be seen in subsequent polls. For now, more than 70 percent of people in the USA believe either that God created us humans or that God has guided our evolution (Figure 1.5).

One might wonder why about three out of four people in one of the most advanced countries in science and technology in the world think that God either created humans or has driven our evolution. There are several facts to consider. For instance, the conclusion from the 2017 survey is that people with higher education are more likely to accept the evolutionary explanation for human origins and less likely to support "creationism." In particular, among people with postgraduate education, 21 percent accepted "creationism" and 31 percent accepted "evolution," whereas this was the case for 48 percent and 12 percent, respectively, of people with high-school education or less. However, the Gallup report noted that "even among adults with a college degree or postgraduate education, more believe God had a role in evolution than say evolution occurred without God." Many would claim that higher levels of education might make one less likely to accept supernatural explanations and more likely to accept natural ones. However, there may be other reasons that about three out of four people in the USA currently seem to prefer responses that implicate God in human origins.

Let me explain. In addition to the Gallup poll, the Pew Research Center has also conducted surveys in the USA about evolution. Let us compare the Gallup and Pew surveys conducted in the USA between 2008 and 2014. I have arbitrarily put together surveys conducted in different years, but my aim is not to compare those but simply to show trends across time. Table 1.1 presents the exact wording of the choices made available to participants in the Pew and Gallup surveys. The findings of these surveys are presented in Figure 1.6.

Even though both the Gallup and Pew polls practically provided participants with the same three options ("evolution," "theistic evolution," and "creationism"), the results in Figure 1.6 indicate two quite different situations. Assuming that all surveys involved a representative sample of the US population, one should not expect significant differences in participants' responses to the two polls. However, this is not the case. On the one hand, twice as many participants selected "evolution" in the Pew polls compared to the Gallup polls. On the other hand, more participants selected "theistic evolution" and "creationism" in the Gallup poll compared to the Pew poll. But how is this

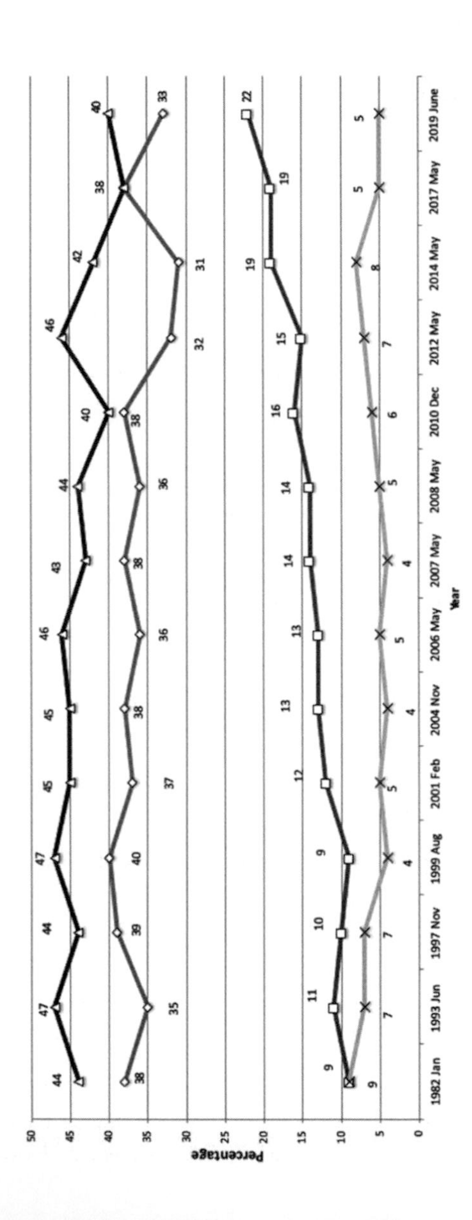

Figure 1.5 Public acceptance of evolution in the USA during a period of 37 years.

Type of question	Gallup	Pew
"Evolution"	Human beings have developed over millions of years from less advanced forms of life, but God had no part in this process.	Humans and other living things have evolved due to natural processes such as natural selection.
"Theistic evolution"	Human beings have developed over millions of years from less advanced forms of life, but God guided this process.	A supreme being guided the evolution of living things for the purpose of creating humans and other life in the form it exists today.
"Creationism"	God created human beings pretty much in their present form at one time within the last 10,000 years or so.	Humans and other living things have existed in their present form since the beginning of time.

Table 1.1 The choices given to participants in the question related to evolution in the Gallup and Pew polls until 2014

possible? If the differences are not due to sample sizes or a biased sampling of participants, what could then account for them? There are actually two important issues, one conceptual and the other methodological.

The conceptual issue relates to the content of the phrases. Look carefully at Table 1.1 and compare the phrases therein. Do you see any difference? Well, the word "God" appears in all Gallup phrases but in none of the Pew ones; the term "supreme being" only appears in one of those. Demonstrating an effect of such differences in wording on participants' responses would require an empirical study in which some participants from the same samples and responding under the same conditions would be given the Gallup phrases and others would be given the Pew phrases. Until this is done, I cannot establish that the differences in phrasing have an effect. However, the

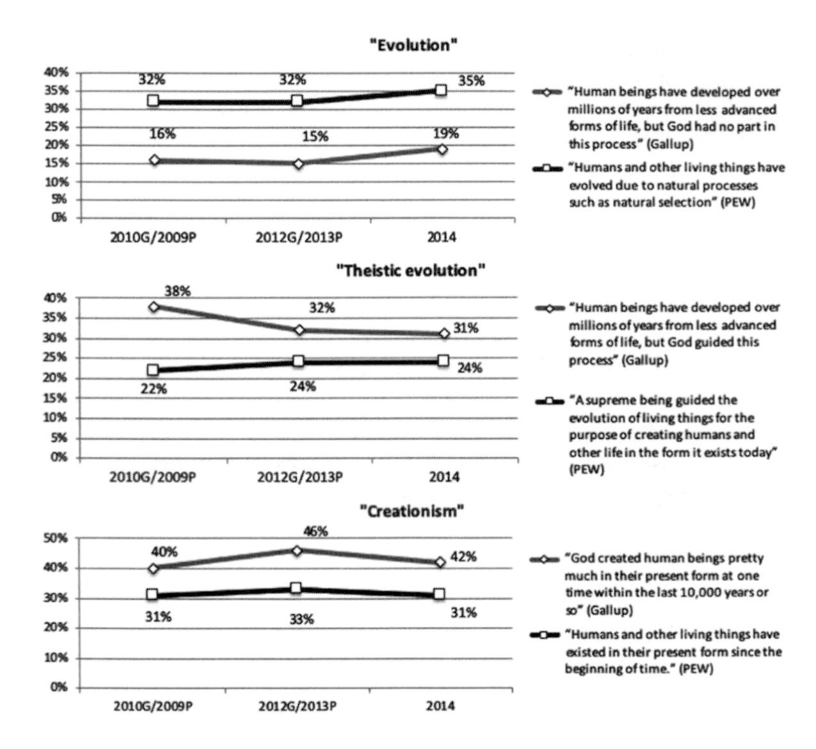

Figure 1.6 The percentages of participants of the Gallup and Pew surveys accepting "evolution," "theistic evolution," and "creationism."

different results in the Gallup and Pew polls make plausible the assumption that this could be the case, and this is something worth investigating.

The methodological issue stems from the fact that, whereas all three statements were given to Gallup participants at the same time, this was not the case for the Pew participants. There, in contrast, participants were first asked which of the following phrases comes closer to their own view: "Humans and other living things have evolved over time" or "Humans and other living things have existed in their present form since the beginning of time." Then, those who chose the first phrase were also asked if they thought that "Humans and other living things have evolved due to natural processes such as natural selection" or "A supreme being guided the evolution of living things for the purpose of creating humans and other life in the form it exists today." But why might receiving all three options together or in two consecutive steps have an effect on participants' responses?

The researchers at the Pew Research Center have actually tested this in a recent survey (April 23 to May 6, 2018). Participants were randomly assigned to one of two conditions: Half of them were asked a single question about evolution and were given three options: "Humans have evolved over time due to processes such as natural selection; God or a higher power had no role in this process"; "Humans have evolved over time due to processes that were guided or allowed by God or a higher power"; or "Humans have existed in their present form since the beginning of time." The other half of the participants were asked about evolution in a two-step process: They were first asked if they thought that "Humans have evolved over time" or if "Humans have existed in their present form since the beginning of time." Then, those who agreed with the idea of evolution were asked a second question, whether they thought that "Humans have evolved over time due to processes such as natural selection; God or a higher power had no role in this process," or that "Humans have evolved over time due to processes that were guided or allowed by God or a higher power."

The results were very interesting. Among those asked the questions in two steps, 40 percent chose "evolution," 27 percent chose "theistic evolution," and 31 percent chose "creationism"; in contrast, among the participants who were given all questions at the same time, only 18 percent chose "creationism," whereas 33 percent chose "evolution" and 48 percent chose "theistic evolution." The Pew researchers thus concluded that testing multiple ways of asking about evolution is necessary and important. They explained the higher acceptance of evolution in the one-step approach as being due to the fact that respondents were not put in the dilemma of creation or evolution, but could choose among all three options right from the start.

The one-step approach is actually the one that Gallup had been using over the years. It is then interesting to compare the most recent Gallup (June 3–16, 2019) and Pew (April 23 to May 6, 2018) polls (Table 1.2) using this approach.

Type of question	Gallup	%	Pew	%
"Evolution"	Human beings have developed over millions of years from less advanced forms of life, but God had no part in this process.	22	Humans have evolved over time due to processes such as natural selection; God or a higher power had no role in this process.	33
"Theistic evolution"	Human beings have developed over millions of years from less advanced forms of life, but God guided this process.	33	Humans have evolved over time due to processes that were guided or allowed by God or a higher power.	48
"Creationism"	God created human beings pretty much in their present form at one time within the last 10,000 years or so.	40	Humans have existed in their present form since the beginning of time.	18

Table 1.2 The choices given to participants in the question related to evolution in the Gallup 2019 and Pew 2018 polls

A first thing to note is that the word "God" that could have made a difference in the Pew–Gallup comparison above now appears in two of the three Pew items as well. However, there are again important differences in the results. Is it because the two surveys took place a year apart? Perhaps. It is noteworthy that "creationism" is the most accepted view in the Gallup poll, whereas it is the least accepted one in the Pew poll. Is this the case because the phrase used in the Pew poll is the only phrase that does not contain the word "God"? It is hard to know. What is important is that there are important methodological and conceptual issues to consider in the interpretation of such surveys.

As the Pew Research Center report concludes: "What may seem like small differences in question wording can have a major impact on survey estimates of the share of the public that believes in a naturalistic account of human development, a creationist view or something in between – an evolutionary process guided or at least allowed by God or a supreme being."

Let us now look at polls at the global level.

Evolution Polls at the Global Level

In 2011, the research company Ipsos conducted a poll for Reuters which involved 18,829 adults from 23 countries. Two of the questions asked were the following:

1. Which of the following sentences best describes your personal beliefs about your God or Supreme Being(s)?
   - I definitely believe in God or a Supreme Being
   - I definitely believe in many Gods or Supreme Beings
   - Sometimes I believe, but sometimes I don't in God/Gods or Supreme Being/Beings
   - I'm not sure if I believe in God/Gods/Supreme Being/Beings
   - I don't believe in God/Gods/Supreme Being/Beings
2. There has been some debate recently about the origins of human beings. Please tell me which of the following is closer to your own point of view:
   - Some people are referred to as "creationists" and believe that human beings were in fact created by a spiritual force such as the God they believe in and do not believe that the origin of man came from evolving from other species such as apes.
   - Some people are referred to as "evolutionists" and believe that human beings were in fact created over a long period of time of evolution growing into fully formed human beings they are today from lower species such as apes.
   - Some people simply don't know what to believe and sometimes agree or disagree with theories and ideas put forward by both creationists and evolutionists.

This poll raises important methodological, conceptual, and inferential issues. First of all, there is the methodological problem that the two questions above are two different kinds of questions. Question 1 is supposed to ask respondents what their beliefs about God are, but the choices given are actually about whether participants believe in God. In contrast, Question 3 is clearly asking people what their beliefs about evolution or creationism are, as people are asked to agree with one of these views or choose the option of being unsure. In short, Question 1 asks people whether they *believe in* God, whereas Question 3 asks them what they *believe about* the origin of humans. But is this an important difference? Yes, it is. Let me use an example to explain. On the one hand, a child might have *beliefs about* Santa Claus, such as that he is nice, generous, has a long beard, and wears red clothes without necessarily *believing in* Santa Claus – that is, without believing that Santa Claus really exists. On the other hand, a child might *believe in* the existence of Santa Claus without holding any specific *beliefs about* him. The case is similar for evolutionary theory. For instance, a biologist might have many *beliefs about* evolutionary theory – that is, understand it and consider it as a useful heuristic for making predictions and conducting experiments. However, that person might not actually *believe in* evolutionary theory – that is, esteeming and valuing it – because it conflicts with her deeply held religious convictions. Conversely, a student might *believe in* evolution – that is, accept that it is true simply because she has heard a prominent scientist saying it is true – without understanding it and thus holding *beliefs about* it.

A second problem is conceptual. Question 3 states that "the origin of man came from evolving from other species such as apes" and that "human beings were in fact created over a long period of time of evolution growing into fully formed human beings they are today from lower species such as apes." The problem here is that these statements might be perceived to imply that humans evolved from the apes that exist today. However, this is not what evolutionary theory suggests; rather, what is suggested is that humans have a common ancestor with these apes, which was probably ape-like, but still different from the apes that exist today. Now, you might wonder, is this a big difference? Ape or ape-like, it is still an ape, isn't it? Well, I think there is. To put it simply: Is it the same to say that you are descended from your first cousin, and that you and your first cousin have a common grandfather and grandmother? The former is impossible, the second is correct. Whether or not the cousin and the grandparents are apes is secondary.

Let us now consider the findings of the Ipsos 2011 study and the respective inferential issues. Figure 1.7 presents how many people in each country

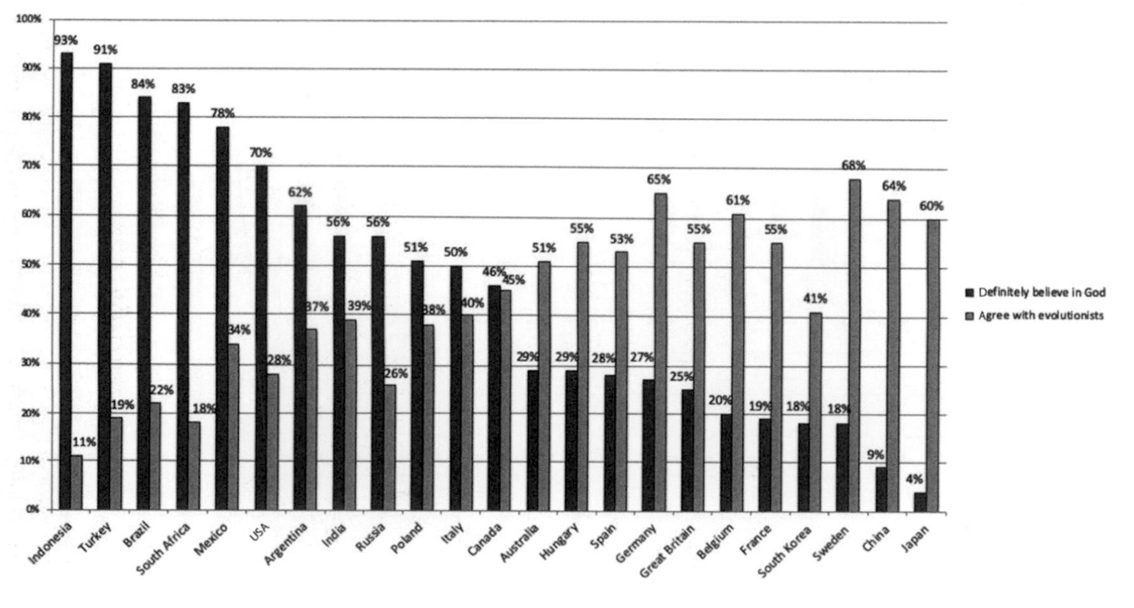

Figure 1.7 The usual image found in polls is that there is a negative correlation between belief in God and acceptance of evolution.

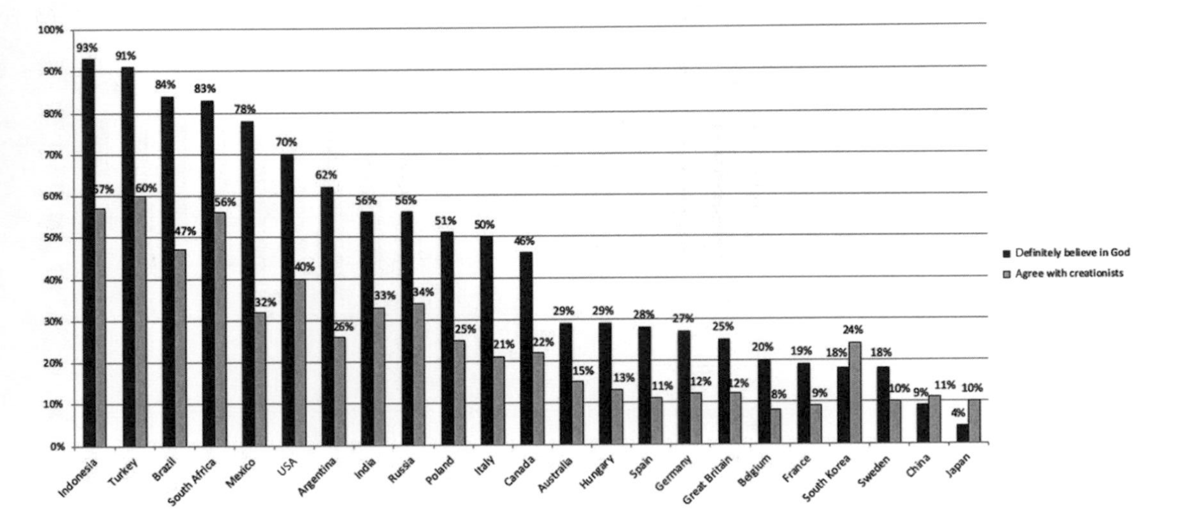

Figure 1.8 There are fewer people who accept creationism than who believe in God; therefore not all religious people are necessarily creationists.

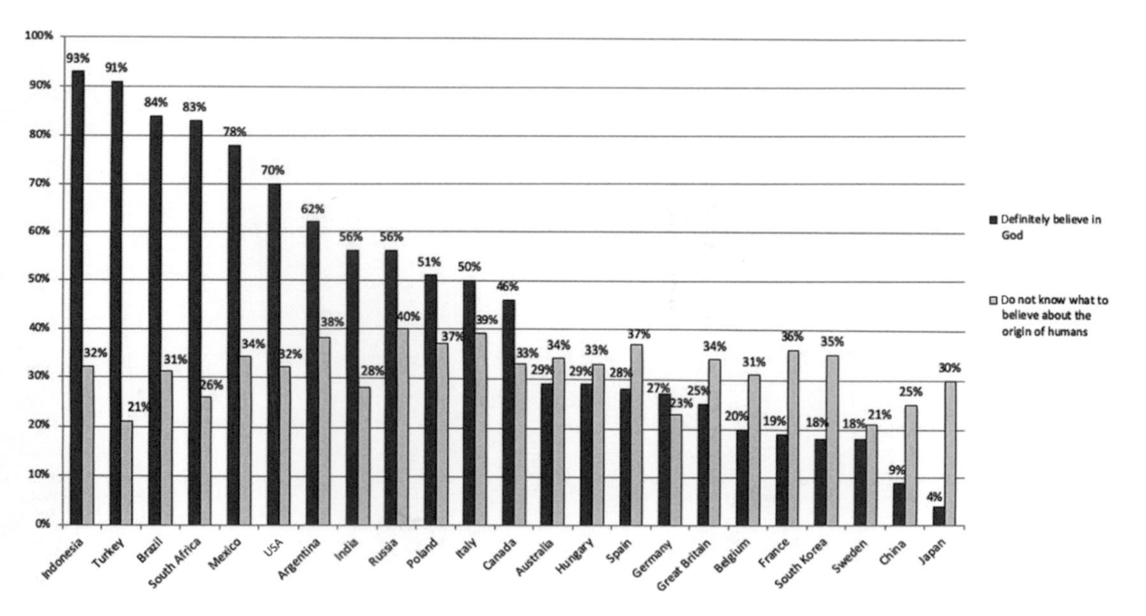

Figure 1.9 Those who were unsure about evolution were more or less equally distributed in "more religious" and "less religious" countries.

definitely believe in God and agree with evolutionists. What emerges is the picture that the more "religious" a country is, the lower is the acceptance of evolution. The immediate inference, then, is that belief in God and acceptance of evolution are not compatible. There is certainly evidence for such a trend. However, given the methodological and conceptual issues discussed above, it is unclear whether this accurately reflects what people actually think. Furthermore, this is not the whole story. On the one hand, if one compares how many people believe in God and how many people agree with "creationists," it becomes clear that in most cases there exist twice as many people who believe in God than there are people who agree with creationists (Figure 1.8). On the other hand, and perhaps most importantly, about one in three people in all countries, both in the "more religious" and in the "less religious" ones, are unsure about what to believe about evolution (Figure 1.9). Therefore, one should go beyond the typical evolution–religion dichotomy to consider those religious people who are not creationists and all those others who are unsure. These are the people who are more likely to understand and accept evolution, and to whom attention in science education and the public understanding of science should be paid.

From all of the above there are two important take-home messages. First, it is far from simple and straightforward to figure out what people think or believe about evolution or religion. Second, even if we are confident that we have found what people believe, there many different kinds of inferences one can draw from the available data. Therefore, caution and thoughtfulness are required before one can reach conclusions. Of course, religious resistance to accepting evolution is a fact, and so it is worth looking into this issue in more detail.

2 Religious Resistance to Accepting Evolution

Purpose and Design in Nature

Contemporary evolutionary theory emerged historically in a Christian religious context. Therefore, in this book I focus on the contrast of evolutionary theory with notions of creation and design in nature within the Christian worldview. Predominant in this case has been the *argument from design*. According to this argument, if nature seems to exhibit design, it is because it is God's creation. Therefore, this design stands as evidence for his existence. A well-known argument of this kind was developed in detail by William Paley, who believed that the complexity and perfection of the natural world, documented by its empirical study, were the most powerful arguments for the existence of God. Paley lived at a time when empiricist philosophers, such as David Hume, had expressed skeptical arguments about this. Hume considered, for instance, that the imperfections that existed in nature did not comply with the idea of design. But Paley argued to the contrary, and tried to show that the existence of God could be confirmed based on the study of nature.

In particular, Paley used the metaphor of the organism as a watch and of God as a watchmaker, according to which a complex structure, like a watch, could not have emerged accidentally but required the existence of a designer-watchmaker. Paley argued that if one came across a stone, one would not wonder about how it came to be there. However, if one came across a watch, one would think differently because of its features. The parts of the watch are formed and adjusted *in order to* produce motion; and the whole system is regulated in such a way *in order to* show what time it is. If the parts of the watch had a different shape or size, or if they were placed in another manner or order, the watch would not be functional. In contrast, the stone has no such features. Therefore, the watch exhibits intelligent and intentional design, which stands as evidence for the existence of a designer.

Paley then made the move to organisms, by arguing that organisms not only seem to be designed as instruments are, but they are also more complex than any human-made object. He then compared the eye with the telescope and concluded that, given the similarities, we cannot conclude that the telescope is designed and the eye is not. The evidence for their close affinity in terms of structure and function, according to Paley, points to the conclusion that there must be some designer for each of them. Furthermore, the eye is adapted to different degrees of light and to observing objects at a variety of distances. These difficulties were not faced by the creator of the telescope. As the creator of the eye not only overcame these difficulties, but did so in many different ways, depending on the way of life of the various organisms, the conclusion is, according to Paley, that the creator of the eye must be more competent than the creator of the telescope. In addition, the creator of the eye made it many times in different ways, in order to make his competence evident to humans.

Bringing everything together, we can summarize Paley's arguments as follows:

Paley's Argument

Complexity indicates design and the existence of a designer.
The more complex an instrument is, the more competent is its designer.
Animal organs are superior and much more complex than human-made instruments.
Therefore, the creator of animal organs is superior and more competent than any human creator.
Therefore, the creator of animal organs is God.

Why did Paley make the inference from complexity to design? Could his religious beliefs be the reason for such thinking? Is it due to religiosity that we intuitively look for purpose and design in nature?

There is a lot of research in psychology that might provide an answer to this question. In one study in the USA it was investigated whether children from Christian fundamentalist school communities expressed more creationist views on the issue of the origin of species than children coming from non-fundamentalist school communities. The participants were divided into three groups based on their age: 5–8 years old, 8–10 years old, and 10–13 years old. All participants seemed to endorse mixed beliefs, with evolution mostly applied to organisms besides humans, for whom creation was instead preferred. Most students from fundamentalist school communities provided creationist explanations for all tasks, independently of their age. On the contrary, students from non-fundamentalist school communities provided explanations that were different in the three age groups. However, 8–10-year-old students of this background provided mostly creationist explanations, which cannot be explained on the basis of religion.

Similar findings were reported by another study that involved US and British students. Based on the assumption that the USA and Britain share many cultural characteristics but differ in religiosity, with the British being less religious than people in the USA, the study aimed to investigate whether there was a difference in the preferences of US and British elementary school students, aged 7–10, for purpose-based explanations. Overall, the explanations of students of both groups were rather similar; they generally preferred purpose-based explanations both for organisms and for non-living natural objects. Therefore, the difference in their religiosity did not seem to have any effect, as the supposedly less-religious British children were as likely to provide purpose-based explanations as the US students. These results support the conclusion that children may be naturally inclined to prefer explanations of nature as an intentionally designed artifact, and that this tendency is not necessarily the result of the religiosity of their social background.

To investigate this further, it was examined whether children's tendency to reason about natural phenomena in terms of a purpose and their intuitions about intelligent design (ID) in nature, whether or not they came from fundamentalist religious backgrounds, were related in any systematic way. British elementary school children (aged 6–10 years old) were given tasks that were expected to document their intuitions about purpose and ID in the context of their explanations for the origins of natural phenomena. It was found that children were most likely to provide purpose-based explanations for artifacts as well as for artifact-like natural objects and animals, but not for natural events. Moreover, children's purpose-based and ID intuitions were found to be interconnected. The results supported the conclusion that there was a systematic connection between children's purpose-based explanations and their intuitions about the non-human intelligent design of nature. Children who provided purpose-based explanations of nature also endorsed the existence of a creator agent, in a manner that might be informed by their understanding of artifacts. However, it was not clear how robust this connection was and if it existed at the pre-school age.

Considering these and other studies, it could then be that cultural factors other than religion, such as our understanding of artifacts, are the reason that we tend to perceive purpose and design in nature. Intuitions about purpose and design come first due to a bias to perceive nature as an artifact. There is thus a need to look for a designer, and it is only then that religious belief comes in. This may be called an artifact-thinking argument and has the following structure:

Artifact-Thinking Argument

> Artifacts are designed by competent designers and this is why they have complex structures.
> We observe enormous complexity in nature – in organisms in particular.
> Therefore, organisms must be designed.
> The complexity in organisms is larger than the complexity of artifacts.
> Therefore, the designer of organisms is more competent than the (human) designers of artifacts, and this could only be God.

Paley's argument, outlined above, has more or less the same structure.

What I am arguing for here is that people make two distinct inferences. The first one is the inference from complexity to design; the second is the inference from design to a designer. Psychologist Deborah Kelemen has suggested that because we grow up surrounded by artifacts and we become familiar with their intentional use from very early in our lives, we may extend our artifact thinking to nature and so come to intuitively believe that organisms, and even non-living natural objects, have also been intentionally designed for some purpose. Therefore, it does not seem to be the case that religious belief makes us think in terms of design. Rather, we are psychologically inclined to intuitively perceive purpose and design in nature, and so we look for a designer; this is what makes religion intuitive. This tendency also makes evolution seem counterintuitive. Let us see why.

Evolution and Worldviews: Perceived Conflicts

Richard Dawkins famously provided a natural equivalent to Paley's divine designer. The argument is simple and shared by many biologists nowadays. There is design in nature, but it is natural; it is neither purposeful nor intentional. There is a designer in nature, but it is a blind and unconscious one. It is natural selection: the blind watchmaker. Dawkins' argument can be summarized as follows:

Dawkins' Argument

> Artifacts are designed by competent human designers and this is why they have complex structures.
>
> We observe enormous complexity in nature, in organisms in particular, which is larger than the complexity of artifacts.
>
> Therefore, organisms must be designed.
>
> The complexity we observe in nature, and in organisms in particular, is the outcome of natural processes, which are unconscious and automatic.
>
> Therefore, if we need to identify a "designer" of organisms that is more competent than the (human) designers of artifacts, this is natural, mindless, and sightless; it can only be natural selection.

Sounds good, no? But why, then, do many people not accept natural selection as an alternative to a divine designer?

Let me compare the two explanations with an example. Imagine a class of 20 students finishing elementary school and getting ready to start middle school; these students have poor grades so far – their average is 10 out of 20; after six years, at the end of secondary school, the class average has reached 19 out of 20. How is this possible? There are two competing explanations, one according to Paley's argument and the other according to Dawkins' argument. The former presupposes an external agent (like Paley's divine designer) who intervenes, whereas the latter does not require one and does not rely on any intention or purpose, but results from a process of selection.

Let us consider the first explanation. The director of the middle school, which has high standards, decides that such poor grades are unacceptable for the school. Therefore, she makes appropriate changes to the curriculum and assigns a teacher-mentor to each student. She also provides students with extra courses and extracurricular material so that they study and learn more. In this case, the director acts like Paley's designer. She has a particular purpose, to improve the average grade of this group of students. In order to achieve this, she designs a six-year-long pedagogical intervention that focuses on each individual student. During this process, each student improves and gradually gets higher grades. At the end of the six-year process the whole class has improved and achieved an average of 19 out of 20. The director's purpose has then been fulfilled, and this was due to the implementation of her intentional design.

Let us now consider an alternative explanation, that of an unintentional process that does not assume any external intervention. The students with poor grades begin middle school. However, this school expels low-attainment students. As a result, during the subsequent six years a process of differential attainment takes place. Those students from the initial group who tried hard enough and learned more, improved and eventually made it to the next grade; in contrast, those who did not, or could not for whatever reason, try hard enough were eventually expelled from school. At the same time, any new student who joined the class in the intermediate grades had to be a high attainer as well. Consequently, over the years the average grade of the class increased as students with low attainment were expelled and only high achievers from other schools were allowed to take their place. Thus, without any designed intervention but only through a process of differential attainment, the average grade of the class improved.

I am aware that this analogy has an implicit assumption for the second case: that the high standards of the school are not due to the intention of anyone (director, school board, etc.) but that they simply exist. If we can overlook this issue, what we get are two cases that represent the differences between divine creation and evolution by natural processes. In the first case the director (who is the analog of the divine creator) implements design to fulfill her purpose: Students manage to attain high grades and stay at this particular school (which is the analog of the organisms' managing to survive in their environment). The fact that students improve is due to the implementation of this design (which is the analog of the divine design of a benevolent God who acts for the good of organisms and who designs their adaptations). In the second case, no such design exists; rather, a process of differential attainment takes place (the analog of the process of differential survival) and as a result the constitution of the class changes, perhaps dramatically, over the years (which is the analog of evolution by natural selection – a process of change through differential survival). Obviously, it is the process of differential attainment that best represents how evolution in nature takes place.

Here is then where a moral problem arises: It seems that it is intuitively more (morally) acceptable for people to think that someone will take care of the low-attainers and help them improve rather than accept that they will simply be eliminated. However, elimination is what often happens during evolution. Therefore, if we intuitively think of organisms as designed in order to be adapted (or adaptable), it is difficult to accept that adaptations originate from chance variations that produce non-designed characteristics, which might later become prevalent due to natural selection. It is even more difficult to accept that evolution in a population results from the death of individuals that do not pass on their characteristics to the subsequent generations. But this is how evolution occurs. In nature, it is not individuals that evolve by undergoing particular changes; it is populations that evolve because some individuals die out and some others manage to survive and reproduce in particular environments. This idea seems to be hard for many people to accept. But why?

A plausible explanation is that it is very difficult for us to get along well with death, even though it is natural, and in fact the only predictable outcome once we are born. Most of us detest death, especially if we have experienced the death of a beloved person. Indeed, the moral values of most human cultures prescribe that we should not cause death, neither to humans nor to other organisms. I do not think that I need to explain why we try to protect human life. In many cases we try to protect animal life as well; we even take animals into our homes as pets in order to protect them, or we protest for their rights. However, the irony is that we often think about death in a discriminative and inconsistent manner. We may protect and love our favorite dog or cat (or whatever – some people make very strange choices of pets), while we domesticate cattle, sheep, goats, pigs, chickens, and other animals in order to eat their muscles (mostly), which are full of protein. Why do we accept their deaths as necessary? To make things worse, we consider someone who kills several people within a single day as a murderer (no question that this person is one), but we tend to consider someone who is doing the same during a war as a hero. Are human lives less important if there is a war going on? I do not really want to get into answering these questions. I only want to raise them in order to show that sometimes we (discriminatively and questionably) consider death as necessary or natural, and sometimes we do not. And the idea that death is a driving force of change in evolution seems to be one we dislike.

This might be a consequence of not appropriately distinguishing between what we know and what we believe. We know that death is a fact of life. We do not know what happens after that, even though many people believe in life after death. We observe coincidental death happening around us, and yet some people believe that it has some deep or transcendent purpose. Even when there is no rational explanation of why someone and not someone else dies, some people are ready to explain in fatalistic terms that there was some reason that someone died and someone else did not. Here the problem of inconsistency arises: Why would God allow one person to die but not someone else? Why do some people live short and miserable lives, like children born in various regions of sub-Saharan Africa, who may die of famine before they grow up enough to die of AIDS, whereas other people live long and wealthy lives in Western countries? A well-known reply is that this may be God's will. He may have some reason that some people die younger than others; that some people live a happy and long life whereas others live miserably and die young. This is called *theodicy*: God supposedly delivers justice as he wishes. But then this assumes that there exist at least two kinds of people: those who have to suffer or die young, and those who do not. Why is that?

To answer this question, it is useful to consider one asked by philosopher David Hume:

> Is he [God] willing to prevent evil, but not able?
> Then is he impotent.
> Is he able, but not willing?
> Then is he malevolent.
> Is he both able and willing?
> Whence then is evil?

If we consider any death caused by anything other than natural causes (e.g., old age) as death caused by some evil power, and call it *unnatural*, then in paraphrasing Hume, we might ask:

> Is God willing to prevent unnatural death, but not able?
> Then is he impotent.
> Is he able, but not willing?
> Then is he malevolent.
> Is he both able and willing?
> Whence then is unnatural death?

The problem here is that any attempt to answer these questions would rather be based on belief, and not on the empirical evidence on which science relies. Even if we assume that God exists, no human has any privileged access to God's mind. But still, many people think they *know* the answers to these questions; however, it could be the case that they simply *believe* they do. We do not know why humans and other organisms die unnaturally; we may of course believe that there is a divine cause for this or that life is purposeless. This is one of the major obstacles in really understanding what is going on around us. This is why it is important to distinguish between what one knows and what one believes. But before addressing this issue, let us first see what scientists think about religion.

Evolution and Religion: Scientists' Views

Being a scientist does not necessarily entail anything about one's religious views. However, it is certainly true that scientists' religious views are different than those of non-scientists. According to a survey conducted by the Pew Research Center between April 23 and May 6, 2018, 33 percent of US adults believe that humans evolved due to processes like natural selection with no involvement by God or a higher power, 48 percent believe that human evolution took place guided by God, and 18 percent believe that humans have always existed in their present form, and thus reject evolution. The situation is very different for scientists, such as those connected to the American Association for the Advancement of Science, among whom 98 percent believe that humans have evolved over time. This may not seem surprising, and indeed the same study found that 76 percent of US adults are aware that most scientists accept evolution. But what does this entail for scientists' religious views?

A systematic study looked into what scientists actually think about religion. The researchers studied a sample of 1646 natural and social scientists from 21 "elite" universities in the USA, interviewing 275 of them. The main assumption was that these "elite" universities are more likely to have an impact on the pursuit of knowledge; it should be noted, though, that the sample mostly included professors from the northeast and west coast of the USA and not from the middle and southern states, where more people are highly religious. Approximately 53 percent of the scientists surveyed stated that they had no religious affiliation, whereas approximately 47 percent of them claimed they had one. Another study surveyed more than 20,000 scientists from the USA, the UK, France, Italy, Turkey, India, Hong Kong, and Taiwan, conducting in-depth interviews with over 600 of them. The researchers reached the following conclusions: There are more religious scientists than commonly thought; religion and science overlap in scientific work; even atheist scientists see spirituality in science; and the idea that religion and science must conflict is primarily an invention of the West. The take-home message of these studies is that scientists are not by default irreligious and that they have a variety of attitudes toward religion (what exactly being "religious" means is a question that does not need to concern us here).

To further show that scientists who accept the fact of evolution may hold very different religious views and may draw very different philosophical conclusions, I outline the views of three evolutionary biologists: Richard Dawkins from Oxford University, Simon Conway Morris from the University of Cambridge, and the late Stephen Jay Gould, who was at Harvard University. While all three of them are proponents of evolution, they disagree on how evolution actually proceeds. Most importantly, they disagree on the implications that evolutionary theory has for our understanding of life and the world. As they have all written books in which they make these views explicit, in this section I attempt an analysis of these in order to explore whether or not they conflate what they actually know with what they believe. These scientists are representatives of three distinct views: atheism on the one side (Dawkins), religiosity on the other (Conway Morris), and agnosticism as an intermediate position (Gould). However, I must note that there in fact exists a continuum of different views that scientists hold.

Richard Dawkins is a well-known atheist. Right from the start of his 2006 book *The God Delusion* he notes that "Being an atheist is nothing to be apologetic about. On the contrary, it is something to be proud of, standing tall to face the far horizon, for atheism nearly always indicates a healthy independence of mind and, indeed, a healthy mind." Dawkins actually considers religious devotion as evidence of unhealthiness:

> You say you have experienced God directly? Well, some people have experienced a pink elephant, but that probably doesn't impress you. Peter Sutcliffe, the Yorkshire Ripper, distinctly heard the voice of Jesus telling him to kill women, and he was locked up for life. George W. Bush says that God told him to invade Iraq (a pity God didn't vouchsafe him a revelation that there were no weapons of mass destruction). Individuals in asylums think they are Napoleon or Charlie Chaplin, or that the entire world is conspiring against them, or that they can broadcast their thoughts into other people's heads. We humour them but don't take their internally revealed beliefs seriously, mostly because not many people share them. Religious experiences are different only in that the people who claim them are numerous.

For Dawkins there is a subjective attitude toward religion. Although the claims of religious people are as irrational as those of mad people, we tolerate religious beliefs because so many people share them.

Dawkins suggests that we are prone to accepting the illusion of design in nature as true and this is why we turn to religion, and think that only God can have created the world around us. According to him, it is an illusion to see design in nature and it is a delusion to attribute this design to God, as he is improbable. Natural selection, on the other hand, is a more probable – and thus a more plausible – alternative. Dawkins also considers the idea of an intelligent designer as a self-defeating one, because he seems to be sure that eventually humans will come up with a cosmological alternative as good as the biological one developed by Darwin. What is most crucial is that Dawkins believes that the question of the existence of God is a scientific one, which implies that it is a question that one day will be answered based on empirical data.

Simon Conway Morris is at the other extreme, but without being as explicit as Dawkins. His response to such arguments is that:

> First, we need to recall the limits of science. It is no bad thing to remind ourselves of our finitude, and of those things we might never know ... Second, for all its objectivity science, by definition, is a human construct,

> and offers no promise of final answers. We should, however, remind ourselves that we live in a Universe that seems strangely well suited for us . . . The idea of a universe suitable for us is, of course, encapsulated in the various anthropic principles. These come in several flavours, but they all remind us that the physical world has many properties necessary for the emergence of life.

Conway Morris suggests that there are questions that cannot be answered by science, and theology might give appropriate insights instead. He implies that there are no empirical grounds on which we can seek answers to the questions about the existence of God. This is one major difference from Dawkins. The other major difference is that Conway Morris seems to be convinced that God exists, because particular features of life on Earth indicate the influence of more than just natural processes: the emergence of workable and adaptive options, the emergence of complexity through the reuse of extant matter, the convergence of features despite the enormous diversity, and the inevitability of sentient life. He implies that these characteristics of life are not accidental and point to factors beyond nature (and thus outside the realm of science). Thus, he suggests that despite its achievements, science alone cannot guide us in our quest to explore nature and understand life. Theology has much to add to this quest, and thus we should aim at unifying these two approaches. What we observe around us cannot be explained solely in terms of scientific inquiry; there is more to it. And although it does not prove the existence of God, we must nevertheless be able to see its value.

There is an interesting contrast so far. For Dawkins, religiosity is evidence of irrationality, if not stupidity. For Conway Morris it is evidence of open-mindedness and thoughtfulness. For Dawkins it is almost certain that God does not exist, and this will eventually be shown by science. For Conway Morris it is almost certain that God exists, but there is no need to prove it; there is abundant evidence around us that should make us undertake scientific and theological quests simultaneously. Stephen Jay Gould expressed an intermediate view:

> I do not see how science and religion could be unified, or even synthesized, under any common scheme of explanation or analysis; but I also do not understand why the two enterprises should experience any conflict. Science tries to document the factual character of the

natural world ... Religion on the other hand, operates in the equally important, but utterly different, realm of human purposes, meanings and values.

For Gould, religion is valuable and can coexist with science. It does not add to the aims of science, but nevertheless has much to contribute to human life. Whether God exists or not is irrelevant, and actually is a question that we cannot answer; but religion is important, this notwithstanding. Gould noted that science and religion occupy two distinct non-overlapping domains, or magisteria, which are equally worthy for any complete human life. In this view, science and religion are entirely distinct and non-overlapping domains that address different questions by providing answers that independently contribute to our understanding of the world. Science and religion can neither be unified nor be in conflict, because they are so different that there is no point in even trying to compare them to each other. They nevertheless can communicate, but without falling into each other's realm. For Gould, science and religion both contribute independently to the fullness of life. The views of Dawkins, Gould, and Conway Morris are summarized in Table 2.1.

	Dawkins (atheism)	**Gould (agnosticism)**	**Conway Morris (religiosity)**
God's existence	Improbable	Unknowable	Probable
Question about God's existence	Scientific	Cannot be answered	Theological
Status of religion for humans	Unimportant and unnecessary, if not harmful	Important	Necessary
Relation between religion and science	Conflict	Coexistence	Unification

Table 2.1 An overview of the views of Dawkins, Gould, and Conway Morris about religion

What is very important and clear so far is that these three scientists have entirely different views about religion. The question that is of interest to us is whether their views are based on actual knowledge or on belief. These three evolutionary biologists all accept the fact of evolution, although they have disagreed about its details. Disagreement among scientists is plausible because it is possible to draw different conclusions from scientific data. But in such cases, there are quite objective grounds – raw data obtained from the study of the natural world – on which conclusions are based. Thus, other scientists can offer constructive criticism or just comment on the different conclusions by studying the available data themselves. But which are the grounds for comparing views about God and religion? I do not think that there are any objective grounds; these views are highly idiosyncratic and subjective. Here is, then, what I think is a major issue in the debates about evolution: People often do not distinguish between what they know and what they believe; this is the distinction to which we now turn.

Distinguishing Between Knowing and Believing

Perhaps it is easy to confuse what we know with what we believe, as we may use the two verbs in the same sense. For example, I may say that I *believe* that my friends love me, while in fact I *know* they do because of their attitude (they visit me often and they always seem happy to spend time with me). In contrast, I may say that I *know* that my friends love me while in fact I simply *believe* this because I have no reason to think otherwise (they may dislike me, but nobody has ever told me that, and I am not concerned that they do not visit me often). In another example, I may say that I *know* that my wife is cooking my favorite meal while in fact I *believe* she is because I just smelled it (but she may not be doing the cooking; someone else may be doing it). Or I may say that I *believe* that my wife is cooking my favorite meal when I smell it, although I actually *know* that she is because I just saw her doing it. The distinction I want to make is between when I believe and when I know. When I see my wife doing the cooking, I have justification to say that I *know* she is doing it. But when I have no justification to say that it is her, and not someone else – such as one of our children – who is doing the cooking, then I merely *believe* that she is doing it just because she usually does all the cooking at home (I am not fond of cooking, by the way).

The implication from this example is that *knowing* requires something more than *believing*. In epistemology, knowledge is considered to be a justified belief. However, there exist different kinds of justification, so that not all justified beliefs are equivalent to knowledge. Knowledge requires well-grounded beliefs. Here is an example: You may have good reasons to anticipate that you will soon be promoted because you believe that your boss thinks very highly of you. Over the years you have worked hard and you have contributed a lot to the success of your company. So, you have very good reasons for anticipating that you will be promoted, because you believe that your boss appreciates what you are doing and that you will be rewarded for that. In this case, you *justifiably believe* that you will get a promotion. But then, this may not be what your boss thinks, and so you never get the promotion. It is, then, possible that one can *justifiably believe* something that is false, whereas *knowing* something entails that it must be true. Therefore, knowledge can be considered to be justified true belief, with justification and evidence being very important. Let us see why.

There are three kinds of ways in which beliefs are grounded: causal, justificational, and epistemic. Each of these is related to a particular kind of question, as presented in Table 2.2. We can distinguish between these grounds by using the example of my wife cooking my favorite meal. Sensing the smell of my favorite meal would be a *causal* ground for believing that my wife is cooking it. She might well be doing this. However, it might also be our son doing the cooking. If my wife had told me that morning that she was going to cook my favorite meal, I would *justifiably believe* that she was cooking it. Again, she might well be doing this. But I might be wrong again, because in the meantime she might have asked our son to cook the meal because something else came up. However, if I went to the kitchen and saw my wife cooking my favorite meal, then I would know that she was indeed doing this. Therefore, there may be causal and justificational grounds for a belief, but some strong kind of evidence is required in order to claim that we *know* something. In the example I used, I actually *knew* when I saw my wife cooking. This presentation is a bit oversimplified, but the main point is that simply having reasons to believe something and/or having a justification to believe something are not enough for having knowledge. Strong evidence is also required that makes a belief true (or at least true enough, as it is usually the case in science).

Grounding of belief	Question	Example with wife and meal
Causal	"Why do you believe that?"	"I smell my favorite meal."
Justificational	"What is your justification for believing that?"	"My wife told me that she would cook it."
Epistemic	"How do you know that?"	"I saw my wife in the kitchen cooking it."

Table 2.2 Grounding of the belief that my wife is cooking my favorite meal, and related questions

Let us consider two examples in order to show why the empirical grounding is crucial for knowledge. Imagine that someone claims to have created the clone of a famous actor. This is theoretically possible, insofar as a person found some biological material (e.g., hair) of the actor, extracted the nuclei of the cells, put a nucleus in an enucleated ovum, and transferred the resulting human embryo to the uterus of a surrogate mother. Whether the baby that will be born is a clone of the famous actor is something that can be put to the test. Experts can extract DNA from the cells of the clone and from the famous actor, sequence the whole genomes of both individuals, and compare them. Given that the two people have a three billion-base genome and that about 100 new mutations can occur during a person's development, experts can make conclusions about how similar the two genomes are. If the differences were less than what one would expect from randomly occurring mutations, one might conclude that the baby is indeed a clone. If the differences were a lot more than what one would expect from randomly occurring mutations, we would know that the supposed clone is not actually one. Scientific knowledge exists only when it is empirically tested in ways like this.

However, it is not always possible to have scientific knowledge through empirical testing. One such case concerns our own unique introspective states. For example, assume that I am currently having a very particular sort of pain sensation in my left leg. Even if I think I know this as surely as I could know anything, this doesn't seem to be something that can be deliberately and independently tested. There is no way that anyone else could have access to the pain sensation I have in order to confirm that I actually have it. However, scientific knowledge most of the time does not have to do with our unique introspective states, but with factual evidence that can be used to test hypotheses. Science asks particular questions about the world, and develops tentative explanations that can be empirically tested. Empirical testing is crucial, because this is how we can find sufficient epistemic grounding for our beliefs. This is perhaps where the main difference between knowledge and justified belief lies. Simply put, a justified belief may turn into knowledge when it is deliberately and independently tested and eventually confirmed.

There are many other beliefs that cannot be tested at all. Someone might claim that God sent him a message that something bad would happen via a sign, such as a depiction of a falling angel being formed by moisture on a window while it was cold outside. This person's causal grounding for the belief that God sent him a message was the sign he saw. His justification for getting this message was the particular depiction he saw (he did not see a dinosaur, but a falling angel), and he therefore considered it as a message from God. Because the angel was a falling one, he considered the message as indicating that something bad would happen. But what are the epistemic grounds for this belief? Unfortunately, there can be no real epistemic grounds for this belief. Even if all people who had seen this sign in the past had eventually experienced something bad, this would still not be actual evidence that the sign was a message from God.

Thinking of the falling angel as an example of religious belief may have connotations of superstition or intellectual naivety. I must admit that many (certainly not all!) religious people I have encountered have exhibited signs of superstition or intellectual naivety. I do not know if religiosity causes one to be superstitious or naive, but it certainly allows one to be so. Nevertheless, to avoid this issue, let me give another example. Let us assume that someone believes that how he leads his life matters to God. Someone who leads a moral life may end up living a very happy and peaceful life. He might thus consider this as a reward from God, who in turn acknowledges that this person never behaved immorally. This could be a sufficient causal grounding for such a belief. If he also knew that all other people leading moral lives were also living happily and peacefully, he would even have justificational grounding for such a belief (although I am inclined to think that anyone reading these lines has already thought that there exist moral people who nevertheless live miserable lives or immoral people who live very happy lives). Even in such a case, it would be very difficult to actually know that how one leads one's life matters to God, because there is no way to test this belief – unless of course someone has some kind of privileged access to God's mind.

This is the critical point. There are many claims that we can test empirically and arrive at knowledge, whereas there are others that cannot be tested empirically and so remain beliefs (sometimes justified, sometimes not). When it comes to science my suggestion is that we should stick with what we actually and currently know, and clearly differentiate it from speculation about what we may one day know or what we may never know. We should always try to distinguish between what we know and what we believe. In the previous section we saw that Dawkins suggested that one day we will know that God does not exist; that Conway Morris implied that one day we will know that God does exist; and that Gould implied that we will never know if God exists. None of these beliefs can currently be put to the test, and so we are free to believe whatever we want. But we need to be responsible enough to distinguish between what we believe and what we actually know. Dawkins', Gould's, and Conway Morris' beliefs about God may be influenced by their own understanding of scientific knowledge, but they do not in any way constitute scientific knowledge themselves. They are philosophical beliefs that cannot be put to the test and cannot be shown to be correct or wrong in scientific terms.

So, what should we do? I think that rationality suggests three steps: First, we need to ask if our beliefs are justified; second, we need to question them and put them to the test; third, we need to accept that some of them constitute knowledge and some do not, and that we need to distinguish between knowing and believing. Scientific knowledge is never absolute, because it is constantly put to the test. The outcome of this process is valuable because it confirms what we know. And if one day it fails to do so, there is no problem, because then we will again know that something we thought we knew was not entirely accurate. Knowledge is important, even if it simply debunks what we thought we knew. What matters for science is to have strong empirical grounding for our beliefs. Beyond this, one should believe what one wants. If we all thought hard about the empirical grounding of our beliefs, we would be able to distinguish between the rather objective scientific knowledge and the rather subjective philosophical views.

This is in my view where the problem of religious resistance to the acceptance of evolution lies, at least in part. People do not easily distinguish between what they know and what they believe. I do not mean to underestimate the religious and emotional issues. However, in my view the conceptual issues are equally important. Perhaps the widespread discussion about evolution and how it relates to religious belief is somehow misleading because it focuses on religious issues – which are one part of the problem – but overlooks the conceptual issues which are another important part. In order to be able to distinguish between what we know and what we believe, we first need to properly understand what we know. As we saw earlier, it may not be the case that people believe in God and then make the inference that God has designed organisms, but rather that people see complexity and design in nature and make the inference to a designer-God. Therefore, it is interesting to explore our conceptual understanding and the related obstacles.

3 Conceptual Obstacles to Understanding Evolution

Concepts and Conceptual Change

Our knowledge and understanding take the form of concepts that are mental representations of the world. Scientific concepts have important representational and heuristic roles in the acquisition and justification of scientific knowledge because they both represent natural entities, properties, and processes, and also make their investigation possible. Concepts should be distinguished from conceptions, the latter being the different meanings of, or meanings associated with, particular concepts. From our early childhood we experientially formulate conceptions of the world that are described as preconceptions. As we grow up, we often assimilate knowledge that further modifies our preconceptions, occasionally turning them into more complex but incorrect conceptions, which are described as misconceptions. Conceptual change is the change of our preconceptions with development and learning; however, that people may restructure or reorganize their conceptions when they acquire new knowledge does not guarantee an accurate understanding of concepts. For conceptual change to occur, individuals must be guided to realize that their prior conceptions are wrong or explanatorily insufficient.

However, this is not easy to achieve because our preconceptions often build on intuitions, which are deeply rooted and strongly held, so that they are not completely overwritten even by expert knowledge. Here is an interesting study as an example of this: Researchers examined whether undergraduate students had any difficulty in classifying as living particular items that children often find difficult to correctly classify. The items included animals, plants, non-moving artifacts (e.g., a towel), non-moving natural objects (e.g., a stone), moving artifacts (e.g., a truck), and moving natural objects (e.g., the water in a river). Participants were asked to state as quickly as possible whether each item was living or non-living. It was found that undergraduate students had difficulties similar to those that young children have in correctly classifying the items presented to them. This shows that childhood intuitions persist into adulthood and are expressed when people are asked to provide answers spontaneously. The same study was repeated with university biology professors to see whether expertise in science has any effect. Interestingly enough, even biology professors had the same difficulties that undergraduates had. Whereas they did better than undergraduates in classifying both animals and plants as living, they did not perform better in their answers for artifacts and non-living natural objects. This also shows that intuitions about what is "living" and "non-living" cannot be completely overwritten, even by expert knowledge.

This is where the problem lies. Strongly held intuitions can generate persistent misconceptions that may serve as conceptual obstacles – conceptions that are strongly held, resistant to change, and thus impede understanding and acquisition of accurate concepts. These, in turn, make scientific theories like evolutionary theory difficult to understand. In this chapter I focus on studies on pre-school and young elementary children and, unless otherwise specified, I use the word *children* to refer to this age range (approximately 4–8 years old). Children of these ages have received little formal training and so their answers often reflect their intuitions. It seems that formal education largely leaves these intuitions unchallenged, and so they may persist into adulthood. Before we get into the details, let us consider a simple example.

When we wake up in the morning, one of our first questions usually is: "Has the Sun risen?" During the day we see the Sun rising and until midday we can even navigate using its position (since we accept that the Sun rises from the east, we know where that is). Then, in the afternoon, we can also see the Sun going down, until it eventually disappears. And so on. Why do we think that the Sun rises from the east and goes down toward the west? Because this is what we observe, and the same pattern is repeated every day. However, anyone who has attended a high-school course on astronomy or who has read any relevant, even non-technical, book on the subject is aware that the previous account for the motion of the Sun is not scientifically accurate. The Sun neither rises nor goes down. It is Earth, our planet, which orbits the Sun, as do all the other planets of our Solar System. The rotation of Earth around its axis and the revolution of Earth around the Sun produce the motion of the Sun relative to our position on Earth that we observe. Even though what we observe can be explained both by a geocentric (Earth-centered) and a heliocentric (Sun-centered) model, our intuitions make us favor the former. However, astronomers have explained that it is Earth that orbits the Sun.

Nicolaus Copernicus questioned the geocentric model of Claudius Ptolemy, according to which the planets encircled a static Earth in the middle of the universe, and attempted to present a new and simpler way of explaining the motions of the planets. What he suggested was that it would be possible to explain some curious features of the planetary motions (such as retrograde motion) if it was assumed that the Sun – and not Earth – was at the center of the universe. In developing his heliocentric model, Copernicus did not introduce any new concepts, but maintained Ptolemy's central concepts. However, Copernicus' heliocentric model required a substantial reorganization of the geocentric model. But what made Copernicus come up with the counterintuitive idea that the Sun does not move around Earth, as we observe every day? The answer is that he considered the broader system, not only Earth and the Sun. Let us see why it is important to consider whole systems (the Solar System) and not just some of their elements (the Sun and Earth alone).

Imagine two people (*X* and *Y*), each of whom drives a car. If they suddenly start moving away from each other, two possible explanations are that either *Y* is static and *X* is moving, or *Y* is moving and *X* is static. Both explanations are equally correct, given what we observe. In order to understand what is happening we need more information about the system. If the system includes a house, and if we examine the movement of *X* and *Y* in relation to it, then we may realize that it is *X* and not *Y* who is moving because *X* moves away from the house, whereas *Y* stays close to it. The point is that scientists go beyond intuitions and simple perception to study phenomena by means of detailed rather than superficial observations, and by taking into account their contexts rather than simply studying them in isolation. In a similar but much more complicated manner, Copernicus not only examined the motion of the Sun relative to Earth, but carefully studied the system that included the other planets. He concluded that a heliocentric model provided a better explanation for the motions of planets compared to a Ptolemaic/geocentric model.

Copernicus thus considered new relations between the planets (e.g., that Earth moves like all the other planets) and rejected the old ones (that Earth is the center of the planetary system). The intuitive belief, firmly held for thousands of years, that Earth lies at the center of the planetary system, was eventually rejected after the crucial contributions of Kepler, Galileo, and Newton. Scientists go beyond intuitive beliefs to study phenomena that may be imperceptible in everyday life. They accumulate data that become evidence for a theory that explains phenomena more effectively than our intuitive theories. But it is here that the problem of public acceptance of science arises. It is not enough to have scientists gather evidence to support their theories; they also need to make the public understand why scientific theories are valid and why they have more explanatory power than our intuitions. Why should I accept the heliocentric model that suggests that Earth orbits the Sun? I do not feel that I am moving at all; I also feel neither any centripetal nor any centrifugal force acting on me. Rather, what I perceive is that I live on a static Earth and I see the Sun and the Moon moving around it. In order to be convinced that this is not the case, someone must convince me that my intuitions are wrong.

Here is where the problem with the acceptance of evolution lies. People tend to intuitively think in particular ways that make the conclusions of scientists about the evolution of life on Earth seem entirely counterintuitive. As we might intuitively think that we live on a static Earth and that the Sun revolves around it, we might also think that organisms are designed and that they have fixed essences. The first intuition about purpose and design in organisms is described as design teleology, whereas the second about essential and unchanging characteristics of organisms is described as psychological essentialism. These are both deeply rooted and strongly held intuitions that generate misconceptions and that make evolutionary theory seem entirely counterintuitive. This is why they are important conceptual obstacles to understanding evolution.

Design Teleology as a Conceptual Obstacle to Understanding Evolution

Why do airplanes have wings? An intuitive answer would be: *in order to fly*. Of course, planes do not fly because of their wings only; they also have powerful engines in order to take-off and maintain flight. How about birds? Why do birds have wings? An intuitive answer would also be: *in order to fly*. Again, birds do not fly because of their wings only; they also have relatively light bones that make take-off and flight possible. However, in both cases wings are perceived to exist in order to contribute to flight. When a characteristic seems to exist in order to fulfill a goal, we may be able to explain its existence by reference to the goal that it serves. Such explanations are called teleological. *Telos* is the Greek word for a final end or goal; therefore, *teleology* is the study of final ends or goals, and the respective explanations are described as *teleological*.

Teleological explanations cannot only be given for parts such as wings, but also for wholes. For instance, if we asked why airplanes exist, a reasonable answer would be: *in order to be used by people for transportation*. How about birds? Could we ask why birds exist? Certainly. But would it be reasonable to answer that birds exist for a purpose? Perhaps. Birds may contribute something important to an equilibrium in their ecosystems, as they function as prey and predators, and one might thus say that they exist in order to do this. Therefore, it seems that we may also give teleological explanations for both artifacts and organisms to questions about wholes, as we did to the questions about parts. However, these explanations are not equivalent for two reasons. First, airplanes are artifacts, whereas birds are organisms. Second, answers about parts and about wholes are different in many respects, because a part may serve a purpose, but a whole may not and may also consist of both purposeful and purposeless parts.

Let us explore the first difference. Airplanes are artifacts, which are objects intentionally designed and created through modification of materials for a purpose. Note: They are not simply *used* for a purpose, but also *designed* for this purpose. Here is an example. One may use a sharp branch from a tree as a tool in order to open holes on the ground. This branch will be an artifact only if it is intentionally modified in order to be sharp and to be used to open holes. If someone found a branch that was accidentally cut to be sharp and used it to open holes, this would certainly be a tool, but not an artifact. Artifacts are intentionally designed for a particular use or purpose, and this is why they have the features (e.g., size/shape) required for it. For instance, airplanes have wings that are proportionally large to their size. A Cessna airplane has smaller wings than an Airbus, and in both cases the wings are long enough to facilitate take-off and flight. No rational engineer would ever design an Airbus with the wings of a Cessna, or vice versa. In addition, there are no airplanes without wings (helicopters and hot-air balloons are aircraft, but they are not airplanes).

To summarize, airplanes are artifacts intentionally designed in order to fly, and they all have the features appropriate for this. Therefore, we can legitimately state that *airplanes have wings in order to fly*.

Is this the case for birds? Is it legitimate to suggest that *birds have wings for flying?* The answer is no. All birds have wings – these are actually their forelimbs – but not all birds use their wings for flying. Eagles fly *because* they have long wings (and light bones, etc.) that make flight possible. But what about penguins? Penguins have relatively small wings for their body size, so it is impossible for them to fly. However, penguins use their wings for swimming, and they can actually swim very fast underwater. Can we then state that penguins have wings for swimming? Yes, we can. Most interestingly, ostriches also have wings but use them neither for flying nor for swimming. And their wings are small in proportion to their body size, at least compared to eagles. And so on. We can thus state that all birds have wings, but not all birds use their wings in order to fly because they are not always of the appropriate size. This happens because birds are not artifacts and their wings were not intentionally designed in order to fly. Birds, like all organisms, have come to possess their characteristics through evolution; they are not the products of intentional design. Actually, the wings of ostriches might stand as evidence against the idea that organisms are intelligently designed. Why would an intelligent designer (especially a divine one) design a big bird with wings that do not help it to fly? A human aircraft designer who designed the airplane on the right of Figure 3.1, which is apparently unable to fly, would be a bad designer, and not one we would think of as intelligent. Ditto for ostriches. Therefore, there is no rational basis for the assumption that organisms are divine artifacts.

Let us now turn to the second difference and to teleological explanations for wholes. Can we explain the existence of airplanes in teleological terms? The answer is definitely yes. Airplanes are artifacts, and they are designed and constructed in order to transport people from place to place. An airplane may be found in a museum, but this is not what it was designed for – that would be an incidental use. If you think carefully, the existence of any whole artifact can be explained in teleological terms: scissors are designed and created in order to cut; nutcrackers are designed and created in order to crack nuts; pencils are designed and created in order to be used for writing; cars are designed and created in order to be used for transportation. And so on. All

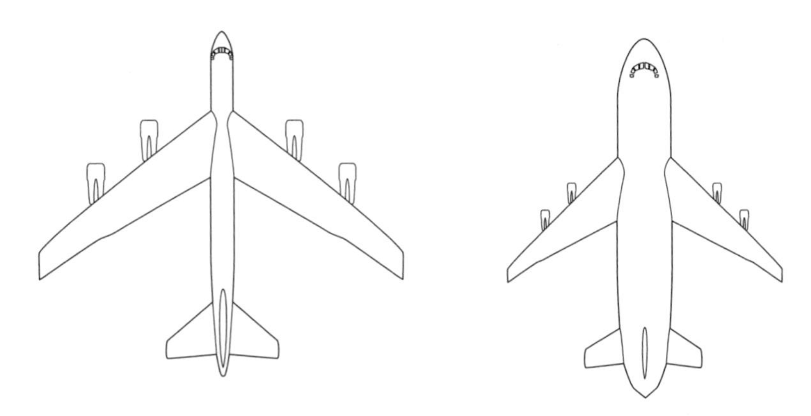

Figure 3.1 How eagles and ostriches would look if they were airplanes. Could you imagine the plane on the right being designed for flying and yet be unable to fly?

artifacts were made for a purpose and have specific features that allow them to (be used to) fulfill this purpose. However, this is not the case for organisms. We know that numerous species that have existed on Earth have become extinct. Assuming that whole organisms were designed and created for some purpose, the question that is really hard to answer is why would anyone create so many species for some purpose and then let them go extinct? Assuming that their supposed role in nature was performed by other organisms as soon as they went extinct, why should have they emerged in the first place? If some organisms serve a purpose in nature, why would anyone create them, let them go extinct, and then be replaced by others? Why not create the latter right from the start? Answers to these questions would be more or less speculative. However, it is more reasonable to think that organisms have evolved and then gone extinct due to natural causes rather than that they emerged for a purpose and were then allowed to become extinct.

Based on the foregoing discussion, we can conclude that the crucial difference between organisms and artifacts is that teleological explanations for artifacts are based on design, whereas teleological explanations for organisms are not. This is especially important when one considers the parts of artifacts and organisms, and the purposes (roles, functions) they seem to serve. Artifacts have particular characteristics *in order to* fulfill some purpose as a consequence of their being designed for it. The wings of airplanes serve their human creators and their intentions. Therefore, in the case of airplanes, *the cause of the existence of the wings of airplanes is humans' intention to fly*. This can be described as *design* teleology. In contrast, the characteristics of organisms do not serve any external purpose because they were not intentionally designed. But what can account for the purposes (roles, functions) that the characteristics of organisms seem to serve? We can legitimately state that an organism's characteristic exists for a purpose when this characteristic exists because of its consequences that contribute to the well-being of its possessor. Regarding birds and their wings, the explanation would therefore be that in the lineage from which birds evolved, those individuals that had wings, or wing-like structures, might have had an advantage in surviving and reproducing compared to those individuals that did not (e.g., because they were more efficient in escaping predators or finding food in remote areas), resulting in natural selection for wings. In this sense, *the cause of the existence of the wings of birds is the survival and reproduction advantage that they conferred to their bearers*. This can be described as *selection* teleology.

With these considerations in mind, let us now look at some studies on teleological explanations. In one study, four- and five-year-old children (and adults) were shown photographs of various organisms, artifacts, and non-living natural objects, and were asked to explain what the objects and their parts were *for*. Participants were explicitly given the option to answer that the objects and/or their parts were not *for* anything. Whereas there was no significant difference between children and adults in providing teleological explanations for the parts of organisms, the parts of artifacts, and whole artifacts, there was a significant difference in that children provided more teleological explanations for whole natural objects, parts of natural objects, and whole organisms compared to adults.

In a second study, it was examined whether children really believed that whole organisms, whole artifacts, and whole non-living natural objects were *made for* something, or whether they thought they could simply perform or be used for certain roles. Children had to choose between statements suggesting that an object was made for something or that it was not made for anything. In this study, there was no difference between children and adults in providing teleological explanations for organisms. In addition, whereas both children and adults generally provided teleological explanations for artifacts, adults provided significantly more teleological explanations for those. Finally, there was a significant difference between children and adults in the teleological explanations they provided for non-living natural objects, with twice as many children providing such explanations compared to adults. Interestingly enough, a third study concluded that both children and adults shared the same notion of function, as they suggested that organisms' parts and artifacts were *for* some particular role.

In an older study, it was investigated whether children preferred teleological explanations for organisms and artifacts in a similar way. Children (second grade) were given two possible explanations for why plants and emeralds were green; one was that being green helps there to be more of them, the other that they were green because they consisted of tiny green parts. The former explanation was given mostly for plants, whereas the latter explanation was given mostly for emeralds. In short, children preferred teleological explanations for organisms than for non-living natural objects.

In an attempt to replicate the findings of that study, another one involved children and adults who were shown pictures of different pairs of organisms and non-living natural objects, and who were then asked "why?" questions about their properties. Participants could choose between two answers for each question, one physical and one teleological. For instance, when shown a picture of an extinct aquatic reptile (*Cryptoclidus*) and a pointy rock found in the same area, and asked why the rock was pointy, participants had to choose between the physical explanation "They were pointy because bits of stuff piled up on top of one another for a long time," or that "They were pointy so that animals wouldn't sit on them and smash them," or that "They were pointy so that animals like *Cryptoclidus* could scratch on them when they got itchy." Adults provided teleological explanations for organisms, but not for non-living natural objects such as pointy rocks. However, children at all grade levels preferred teleological over physical explanations for the properties of non-living natural objects such as rocks and stones. In a second study, an attempt was made to influence children's explanations about non-living natural objects. Thus, children were not only shown the picture of a cloud, but also additional pictures that presented the stages of cloud formation. This was a hint toward physical explanations; however, results indicated that this had no influence as there were no overall differences from the previous study.

This evidence supports the conclusion that children provide teleological explanations for all objects in a non-discriminative manner, drawing on their understanding of artifacts. Assuming that children become familiar with artifacts before they become familiar with organisms and non-living natural objects, it is reasonable to conclude that children extend their intuitive artifact thinking to organisms and to non-living natural objects. However, not all research findings support this. For instance, in another study, four- and five-year-old children asked different types of questions for animals and artifacts. The researchers found that more children asked questions about the functions of artifacts than about the functions of animals. For instance, the question "What does it do?" occurred more often for artifacts than for animals. In addition, children asked what artifacts were designed for or how they worked, but never asked such questions about animals. Such results stand as evidence that children perceive organisms, or at least animals, differently than artifacts.

A relevant important finding is that young children are sensitive to intentionality. In a study, three-year-olds, five-year-olds, and adults were shown representations of objects (drawings and paintings) and actual objects. Both children and adults tended to attribute a function to the item if it was described as intentionally created, rather than if it was described as accidentally created. For instance, when shown a knife-like structure made of Plexiglas that looked very much like a knife, more children called it a knife if it was intentionally created rather than if it was accidentally created. The shape of the object was not the only factor that affected participants' decision to call it a knife; it was also important whether or not it was intentionally made.

In a follow-up study, children were shown the processes through which the items could have been made (intentional and accidental) without being given any further information. For example, a splotch of yellow paint that looked like the Sun was made in one case by accident (and the woman who made it seemed to be disappointed by what she had made), and in another case intentionally (and the woman who made it seemed to be satisfied by what she had made). Children relied mostly on intentionality to name the objects that were being created.

Finally, in another study, three- and four-year-old children were asked questions about novel artifacts. Then, some children were shown a function that plausibly accounted for the structural features of the object, whereas others were shown an implausible function. Children given plausible functions were more satisfied than those given implausible functions, because the latter asked more questions about function. This suggests that children seem to think in terms of intentional design when they think about functions, and that they consider the design functions to be the true functions of artifacts.

The studies reported previously have two common features that must be noted. First, in all studies children generally provided teleological explanations for organisms and artifacts. Second, even if children perceive animals as being different from artifacts, this does not entail that they perceive animal parts differently from artifact parts. For instance, in one of the studies reported above, children's questions about function were more frequent for animal parts than for whole animals, and overall the number of questions about parts was similar for organisms and artifacts. Another study also reported that four- and five-year-olds provided teleological explanations for both animal and artifact parts, while they also realized that parts of organisms are more likely to have some use or function compared to whole organisms. In my view, this is very important. It may be the case that children tend to explain the existence of specific parts of both organisms and artifacts in similar terms, even if their overall perception of artifacts and organisms is different. But is it then knowledge about organisms or knowledge about artifacts that influences the other? There is no simple answer to this question, and there actually exists conflicting evidence.

I personally find very plausible the idea that knowledge about artifacts influences knowledge about organisms. Let me explain why. From very early in our lives, most of the objects that we encounter are artifacts and we learn that they exist in order to achieve some goal. Just think how many artifacts were around you when you were an infant; you may have spent most of your first 2–3 years of life inside your home surrounded by toys, feeding bottles, a crib, dummies, chairs, tables, sofas, spoons, plates, electronic devices, and numerous other artifacts. How many animals did you encounter as an infant? Did you even pay any attention to non-living objects such as clouds or rocks? Even if you grew up in the countryside and your parents were farmers, the animals and the non-living natural objects that you encountered were probably much fewer in number and with less variation compared to the artifacts that you were familiar with at a young age. Thus, it may be the case that in the absence of alternative explanations, children intuitively draw on their early awareness, understanding, and knowledge of the intentional creation and use of artifacts and eventually conclude that organisms, like artifacts, also have parts that exist in order to be used for some purpose.

Based on all the above, I suggest that design-based, artifact-like, teleological thinking about organisms *can* be an important conceptual obstacle to understanding evolution. If children explain the existence of the wings of birds in the same way that they explain the existence of the wings of airplanes, it is important for them to realize from as young an age as possible the difference between birds and airplanes, and more broadly between organisms and artifacts. Even if it is eventually shown that teleological thinking about organisms does not stem from an understanding of artifacts, science education and public communication about the theory of evolution must clearly highlight the differences between organisms and artifacts, and address people's perceptions of purposeful design in nature. I return to suggestions about how this might be accomplished in the final section of this chapter. We now turn to another important obstacle that is also relevant to artifacts: psychological essentialism.

Psychological Essentialism as a Conceptual Obstacle to Understanding Evolution

Both birds and airplanes have wings. Thus, we might plausibly subsume these under the same category: *objects with wings*. In a similar manner, we might also think of an elephant and a car sharing an important similarity that brings them under the same category: *objects without wings*. But we do not usually do this, because despite such similarities (having/not having wings), several other important differences exist between birds and airplanes, as well as between elephants and cars. Birds and elephants reproduce and develop, whereas airplanes and cars do not. Furthermore, as explained in the previous section, airplanes and cars are intentionally designed for a purpose, whereas birds and elephants are not. Consequently, we would rather classify birds and elephants as organisms and airplanes and cars as artifacts. And we might distinguish between them due to some characteristic properties they have, which we might consider to constitute their essence: a set of properties that all members of the kind must have, and the combination of which only members of the kind do, in fact, have.

What could the essence of artifacts be? Could it be their appearance? Think of a kitchen fork. A kitchen fork has a prong. Can we claim that whatever has a prong is a kitchen fork? No, because pitchforks also have prongs. Does it make a difference whether the prong is made up of three or four tines or more? No, because although kitchen forks usually have four tines, those for babies may have fewer. And although a pitchfork may look like a kitchen fork, we would not call it one. The reason is that we know that a kitchen fork is an object we use in order to eat, whereas a pitchfork is an object we use in order to clean our yard.

In a similar manner, we can certainly distinguish between a knife and a sword, and we would never ask for a sword to cut the bread for dinner. How about chairs? We cannot define a chair as an object that has four legs, because tables also have four legs. And if we also think that there exist kitchen chairs, office chairs, wheelchairs, arm chairs, etc., we can realize that there is no single way we can provide a general description of chairs based on their appearance. What, then, is the essence of artifacts? It seems reasonable to suggest that the essence of artifacts is determined by their intended use – in other words, what they were made for, because this is what makes them distinct from one another.

As already explained in the previous section, artifacts are by definition objects created for an intended use. Thus, we distinguish between a kitchen fork and a pitchfork, or between a knife and a sword on the basis of their intended use, and not of their shape – it may be the case that a knife and a sword differ only in size, while they may be quite similar otherwise. We are also able to identify a chair from a table, although they both have four legs because of their intended use. We would not normally sit on a table and we would not put our meal on a chair in order to have lunch. And if we used two chairs for dinner, one to sit on and another to put our meal on instead of a table, the latter would still be a chair even though we used it as a table. Similarly, we can distinguish between a football, a basketball, and a volleyball; they not only have different colors, but also different sizes and weights that are appropriate for the respective games. We might use a volleyball to play football, but it would still be a volleyball. How about using a chair and a ball and other materials to create a scarecrow? We might never use that chair again to sit on or the ball to play; nevertheless, the chair would still be a chair and the ball would still be a ball. In short, the identity of artifacts is largely determined by their intended use because this in turn determines their particular features. Artifacts are designed for an intended use that must be served by their features, and so the latter reflect this use.

One might argue at this point that artifacts do change because they may rust, rot, decay, or more generally undergo changes in their appearance or internal structure. However, even then they still retain some properties relevant to their intended use. A rusted chair is still a chair, even if we never sit on it again. Now, if we broke the chair into pieces and made a new artifact out of it, say, connect the legs to each other to make a billiard stick, most people might recognize that it is not a genuine billiard stick but one made of chair legs. But if we processed the chair legs entirely, then we would have made a new artifact because of our actions that gave to the material a new intended use. In other words, the essence of artifacts would change only when the change in appearance was due to some human intervention and processing with a new intended use in mind. Otherwise, any change that artifacts would undergo would be superficial; the initial parts would be there, even if they were somehow different from their initial state. It is in this sense that artifact essences are considered to be fixed. The initial intended use is evident in their structure until they are consciously transformed with a new intended use in the mind of the human who makes the transformation.

What is the essence of organisms? We usually perceive particular properties as essential for organisms, i.e., characteristic of them. Organisms reproduce, develop, respire, digest food, excrete waste products of metabolism, react to stimuli, etc. We are also able to distinguish between particular types of organisms. It is often easy to distinguish between tigers and lions because the former have stripes whereas the latter do not. It is also easy to distinguish between rhinoceroses and hippopotami because the former have a horn whereas the latter do not. But are these "essential" characteristics? The answer is no. Otherwise, we might claim that all animals with stripes are tigers (but of course they are not; zebras also have stripes), as well as that all animals without stripes are lions (but many animals, such as horses, do not have stripes). Or we might claim that all animals with horns are rhinoceroses (but what about bulls, goats, reindeer, etc.?), as well as that all animals without horns are hippopotami (they are not of course – horses, donkeys, pigs, and many other animals do not have horns). How about using more specific characteristics, such as those used in taxonomy? Could we distinguish between birds and mammals on the basis that birds have feathers whereas mammals do not, and that mammals have hair (or at least hair follicles) whereas birds do not? This is a more reasonable approach and one that actually helps us categorize individual organisms into classes. But this approach may also face problems: How should we classify *Archaeopteryx*? As a bird (because it has feathers) or as a reptile (because it has a dinosaurian skeleton)?

If organisms have essences, these have nothing to do with souls or any other transcendental notions; however, they should be at some deep level as that is by definition what an essence is about. An interesting suggestion is that the essences of organisms are related to their developmental capacities. The characteristics of organisms are the outcome of development on the basis of a particular genetic material expressed under particular environmental conditions. Therefore, their essences are related to the developmental potential of their embryos to produce robust outcomes (an eagle will have wings but a human will not). Artifacts, in contrast, do not exhibit any such developmental capacities. We can therefore state that organisms have developmental essences, which is their potential to develop toward specific forms.

This entails that there are important differences in how the essences of artifacts and organisms may change. Artifact essences are more fixed compared to organisms' essences. A chair is designed to be used for sitting, and the intention of its designer cannot change once it is created. We may adopt a chair for alternative uses – for example, we may use it as a table if we have two chairs but no table. We might also slightly modify a chair to make it look like a small table (by taking away its upper part). More generally, we might slightly modify any chair and co-opt it for several desired uses. But it would still be a chair, albeit a modified one. Artifacts can also evolve, but this is a case of cultural evolution as artifacts are the products of human culture. For instance, think of the primitive tools humans once used for cutting. These were initially stone tools that were hewed simply by being struck against larger stones. Later, these stone tools were replaced by copper or iron ones, which were more efficient cutters. Nowadays an enormous variety of cutting tools exists.

Today's artifacts are the outcome of artificial selection, which is a conscious, intentional process. Over the years, people have been modifying artifacts or creating new ones, and eventually selected to keep creating the ones that were more useful and stopped creating the less useful ones. None of the cars produced in the beginning of the twentieth century are produced by car industries today, as no one would buy them except some romantic car lovers. The cars produced today are much faster, safer, and friendlier to the environment, and these are the outcomes of artificial selection in the car industry over the past 100 or so years. Finally, artifacts do not share a common ancestry. They were developed at different times, under different conditions, and for different purposes. Particular designs were implemented in each case, and the several types of artifacts have had independent origins from the others. Even if some primitive cutting tools evolved to more modern ones, their lineage is entirely independent from that of primitive means of transportation that evolved to contemporary cars. Common ancestry among artifacts is possible, but not obligatory.

Organisms differ from artifacts in all these respects. Organisms can undergo changes in their essences more easily and more drastically than artifacts. We cannot take away the upper part of an animal or plant to make it a different one, nor can we turn a pig into a lion. However, the genetic material of organisms can change, either due to selective breeding or due to mutations – that is, changes in their DNA. These processes involve significant changes or alterations in the genetic material of the organisms, which is part of their developmental essence. Consequently, the essences of organisms, their fundamental developmental properties, are less fixed than those of artifacts. It is exactly this that makes the evolution of organisms possible. With various phenomena that cause genome alterations (mutations, horizontal DNA transfer, genome acquisitions, etc.), and through several processes (natural selection, genetic drift, etc.), individual organisms can change significantly and thus populations can evolve. Evolution is a purposeless, unintentional process that depends on the developmental potential of organisms, as well as on their particular environment. What is also important is that all organisms, contrary to artifacts, share a common ancestry. This is why some fundamental characteristics are common to all organisms. Perhaps the most important ones are that all organisms have DNA as their genetic material, that all organisms consist of cells, and that many of them can undergo developmental changes. Common ancestry among organisms is the outcome of biological evolution – this is not the case for artifacts (Figure 3.2).

There exists an enormous body of research that supports the conclusion that people tend to intuitively provide essentialist explanations for the characteristics of organisms from very early in childhood. Children think that internal

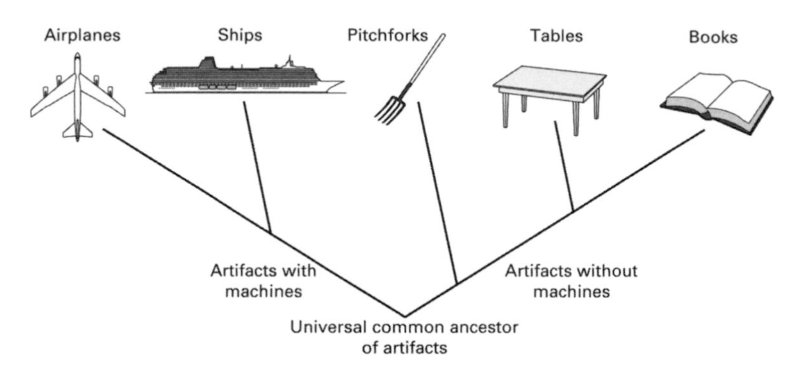

Figure 3.2 An imaginary depiction of several artifacts sharing common ancestry. This depiction is, of course, incorrect as these artifacts have independent origins. They were designed and made independently for different uses. However, do you realize that organisms as diverse in appearance as these artifacts all share a common ancestry and various fundamental common characteristics?

causes are more appropriate for organisms and external causes are more appropriate for artifacts. Interestingly enough, children are not essentialists about artifacts, at least in the way they are for organisms. Essentialism is for them, first and foremost, a characteristic of organisms. This tendency is described as *psychological essentialism*: Certain categories are real rather than human constructions, and they possess an underlying essence, which is responsible for why category members are the way they are and share so many properties.

In one study, four-year-old children were shown pictures of a colorful tropical fish, a gray dolphin, and a gray shark. They were shown the picture of the fish and told that it breathed underwater; they were shown the picture of the dolphin and told that it should pop out of the water in order to breathe. They were then shown the picture of the shark (who looked similar to the dolphin) and asked if it breathed like the fish or like the dolphin. Many children answered that the shark breathed like the fish, and more generally they based their answers on category membership (the tropical fish and the shark were both fishes) and not appearance (the shark and the dolphin were gray and had similar shapes and sizes). This is an important finding that has been replicated with younger children (two-year-olds). It seems that young children with no scientific or other relevant training make inferences about an animal based on another animal that they consider to belong in the same category. In a follow-up study, three- and four-year-olds overall made inferences based on category membership, not external appearance. Among the items shown were an insect (beetle), a leaf, and a leaf-like insect. The last two were both large and green, with striped markings. However, despite the similarities between them, children drew inferences from the leaf-like insect to the beetle because they could identify them both as bugs. They did not draw inferences from the leaf-like insect to the leaf because they noticed the antennae on the former. So, again, children's inferences were based on category membership and the related underlying, non-obvious properties, and not apparent similarity.

In another study it was found that children thought that animals, contrary to artifacts, retain their essential properties despite transformations they might undergo. In particular, the following main types of transformations were shown, with pictures, to children (kindergarten, second grade, and fourth grade): animals into animals (e.g., horse into zebra); plants into plants (e.g., rose into daisy); non-living natural objects into non-living natural objects (e.g., salt into sand); artifacts into artifacts (e.g., bridge into table); animals into plants (e.g., squirrel into moss); machines into animals (e.g., toy mouse into real mouse); and animals into non-living natural objects (e.g., fish into stone). The main aim of this study was to assess how much children were basing their categorization of objects on apparent features. In all cases, children were told that a scientist took the first object of a pair and performed appropriate operations to turn it into the second object of the pair. The results were extremely suggestive: Children of all grade levels answered that kind was preserved in transformations from animal into non-living natural objects (e.g., although a hippopotamus looked like a big rock, it had not turned into one) and animal into plant (e.g., although a squirrel looked like a moss plant, it had not turned into one), despite apparent similarities. They also thought that kind was preserved in machine-to-animal transformations (if an entity had a machine inside, it could not be a real animal – e.g., a toy bird could not turn into a real bird). In contrast, they thought that kind was not preserved in artifact-to-artifact transformations. Fourth-grade children also clearly thought that kind was preserved in animal–animal, plant–plant, and non-living natural object–non-living natural object transformations. However, this was not the case for younger children, especially kindergarten children, who generally thought that kind had changed. Overall, second-grade and especially fourth-grade children thought that kind identity persists despite changes in appearance.

An important question was whether kindergartners in the previous study did not think that identity was maintained because they did not understand the nature of the transformations performed. Thus, a new study was conducted with kindergartners and second-grade and fourth-grade students. Although the same photographs were used – so that the beginning and end-state characteristics were the same for each pair as in the previous study – children were told that a different kind of transformation had taken place. Instead of being told that a scientist made changes in an animal (e.g., that she put black and white stripes on a horse, or that she taught it to run away from people and to live in the wild part of Africa rather than in a stable), children were told that the animal was dressed in a costume or that a superficial transformation had taken place so that it eventually resembled another animal (e.g., that a man was painting black stripes on his white horse every week). One important difference from the previous study was not only that the transformation was more superficial, but also that the children were told that it had to be repeated regularly to ensure the animal would not revert to its initial state. These two new conditions (superficiality of transformation and the need for regular repetition) were expected to make even kindergartners suggest that kind had not changed. As was expected, it was found that even the youngest children denied that any change in kind had taken place. The results for artifacts were similar to those in the previous study. These findings seem to indicate that even pre-school children do not rely on external appearance, but on some deeper properties in order to decide about whether a change in kind has occurred.

If children do not consider external features as characteristic of organisms' identities, do they consider internal parts of organisms as more important? To investigate this, four- and five-year-olds were asked to consider particular transformations during which either internal or external parts were removed. For example, they were asked whether the identity of a dog would change and whether it would still bark and eat dog food if some of its inside parts such as blood and bones were removed, but the outside parts were left intact, or if outside parts such as fur were removed but the inside parts were left intact. The questions also included containers, like refrigerators and jars, the insides of which are not their integral parts. Children answered that the identity of containers would not change if their inside parts were removed. However, in the case of entities for which inside parts are important, such as a dog or a car, the engine of which is more important for its function compared to its paint, children thought that inside parts were significantly more important compared to outside parts. Again, children were found to think that some deeper features/properties are more important than external ones.

An interesting question that comes next is: What do children think about biological transformations that occur in nature? Both evolution and development are processes of biological transformation, but it is the latter that can be easily observed, even by young children. An interesting case is that of some animals whose development includes extensive transformations, known as metamorphoses (e.g., butterflies). Do children think that natural biological transformations lead to identity change? In one study, three- and five-year-old children were initially shown the picture of a caterpillar. Then they were shown another picture identical to the first one, and one that was the same but larger in size, and were asked which one represented the adult form. All of the children chose the larger figure, which shows that they thought that growth does not affect identity. In another task, children were shown a set of pictures – first a picture of a caterpillar and then a smaller picture of the same caterpillar, together with the picture of a moth that was larger but very different from the caterpillar. A significant number of five-year-olds chose the moth as the adult form of the caterpillar. It seems that by the age of five years children realize that organisms can undergo radical changes without a change in their identity.

The main conclusion from this research is that children intuitively think about organisms and artifacts in exactly the opposite way to how they should. They think that organisms, not artifacts, have fixed essences. They perceive organisms as capable of undergoing changes in their external features without undergoing any change in their identity. In contrast, children seem to think of artifacts as undergoing changes in identity when they simply change shape or form. Overall, children perceive that particular properties are characteristic ("essential") to organisms (something inside them) and to artifacts (their intended use), but they consider organisms as having fixed properties that cannot change, and think exactly the opposite about artifacts. Here is then why psychological essentialism, the intuition that organisms have fixed essences and that they do not undergo a change in kind even when their external features change, is a major obstacle to understanding evolution. If children think that organisms have fixed essences and that they cannot change kind, then it is difficult for them to understand the idea of evolution.

However, this is only part of the problem. Transformations such as those presented to children (horse to zebra or caterpillar to moth) are very different from the changes that take place during evolution. Evolution takes place through changes in populations and not in individuals. In a sense, children's essentialist bias is a denial that individual essences undergo change. This is actually consistent with evolutionary theory, which suggests that evolution takes place across generations and within lineages, not during individual lives. These studies reveal not only that children think that organisms cannot change significantly, but also that they do not realize that organisms of the same kind/category may exhibit an enormous amount of variation. Therefore, the major obstacle that essentialism poses to understanding evolution is not only that some "essential" properties are fixed, but also that these are perceived to be identical in the members of the same kind. Evolution requires variation, and psychological essentialism seems to preclude students from thinking about how much the individuals of the same population may vary.

Conceptual Change in Evolution

There are at least four ways in which psychological essentialism may pose obstacles to understanding evolutionary theory: (1) the assumption that categories are stable conflicts with the idea that species can evolve and change over time; (2) the tendency to intensify category boundaries makes it difficult to understand that two species may have a common ancestor; (3) underestimating the amount of variation within a category may make it difficult for people to understand how natural selection, which requires variation, operates; and (4) essentialism reinforces a focus on inherent causes within individuals rather than on the characteristics of a population, leading to a misunderstanding of evolution. Design teleology can also pose obstacles to understanding evolution in several ways. One is that children may generally view natural phenomena as existing for a purpose due to underlying intuitions that make them believe that such phenomena derive from intentional design. Alternatively, children's generalized tendency to ascribe functions to natural entities may result from a basic cognitive mechanism that makes children view entities as made "for" a purpose based on simple cues about functional utility. For all these reasons, evolution is difficult to understand.

Let us now consider how conceptual change in evolution might be achieved. Two main conclusions from the studies from conceptual development research reviewed in the previous sections are that from a very young age children: (1) provide teleological explanations for both organisms and artifacts; and (2) think that organisms have more fixed essences than artifacts, as well as that they do not vary significantly. Therefore, what is necessary for students to understand from as young an age as possible are the major differences between organisms and artifacts. Artifacts are designed and created for an intended use. Consequently, they have parts that serve their intended use, and this may be perceived as their "essence." In contrast, organisms are neither designed nor created for any intended use. If they have some parts that seem to serve some use (which is entirely unintended), these have emerged through a long evolutionary history and may have been maintained through natural selection. The outcome of this history is a specific developmental potential, which could be considered as their "essence." Therefore, artifacts and organisms are "essentially" different.

What is the problem with design teleology? Teleological intuitions, perhaps stemming from an early awareness of intentionality, may make us unconsciously think about the parts of organisms in the same way we think about the parts of artifacts. This does not necessarily mean that we consciously consider organisms as (divine) artifacts. Rather, it may simply mean that we think of both organism and artifact parts in terms of intended uses, because this is what we perceive them to perform. For example, seeing an eagle flying does not necessarily make us think of it as an artifact, as we would do for an airplane. However, it may be the case that seeing an eagle flying by using its wings makes us think of the wings as parts that exist for this particular use, in a similar way that we think of the wings of an airplane as existing for flying. Now, whether or not there is an intentional agent behind this use of parts is a distinct question that comes afterwards. Parts may have a use, independently of whether it is intended by someone (artifacts) or not (organisms). It is the intuitive idea of an intended use, *intended for* some purpose (and perhaps, but not necessarily, *intended by* someone) that makes design teleology an important conceptual obstacle to understanding evolution – in my view, the most important one. The main issue is not whether a part of an organism exists for some purpose, but whether it was intentionally made to fulfill this purpose. In other words, the problem is not teleology per se, but the design assumptions underlying it.

What is the problem with psychological essentialism? There are at least two problems: On the one hand, we may think that change is impossible because organisms are perceived to have fixed essences; on the other hand, the notion of a group of characteristic properties that also determine kind membership does not let us realize how enormous is the variation that exists within each kind. The latter relates to an important linguistic issue, which has to do with how we refer to kinds and their members. Consider the following sentences: (1) *The* eagle is flying by using its wings; (2) *An* eagle is flying by using its wings. Although these sentences seem similar, they nevertheless have different referents. "*The* eagle" implicitly refers to an exemplar, which possesses all those ("essential") properties required in order to classify a bird as an eagle. It may impose the notion of a prototype to which all individuals of the kind must be identical. In contrast, the phrase "*an* eagle" implicitly refers to a particular individual only, which allows the possibility that it may differ from other individuals of the same kind. It seems that we intuitively think that all individuals belonging to the same kind must be "essentially" the same, and thus overlook the enormous within-group variation that actually exists in nature. If all members of a kind had the same fixed "essential" characteristics, only small changes in "non-essential" ones might be possible. But evolution is about substantive changes that take place over time.

Bringing all this together, design teleology and psychological essentialism produce the following (mis)conception: Organisms of a particular kind/category have particular fixed characteristics that exist for some intended use. But this is true only for artifacts, not for organisms. Here is then a very important issue that causes misunderstanding of evolution: *artifact thinking, or (unconsciously) thinking about the parts of organisms as if they had specific intended uses*. Let us now try to describe this teleological–essentialist bias in a single sentence: *The* eagle has wings *in order to* fly. Note that I have put emphasis on the words that represent the essentialist bias (*the*) and the teleological bias (*in order to*). Can we rewrite the sentence to address these biases? Yes. Here it is: Eagles fly *because* they have wings. In this phrase, not only the biases have been eliminated, but there is also a causal connection between the parts (wings) and the use (flying). There is no essentialism because reference is not made to any prototype eagle, but to eagles (plural) that may slightly differ from one another in various characteristics, including the ones we consider distinctive of their kind. Also, there is no teleology because function (flying) comes after structure (wings). Eagles fly because they have the wings that they do, but wings are not always used *for* flying; penguins use them for swimming, whereas ostriches do not use them at all. With these changes, the explanation for the existence of wings has changed, and this is what conceptual change in evolution is about.

I must note at this point that the teleological component of this explanation is rather tricky, and this is why I think that teleology is the most important obstacle to understanding evolution. Let me explain. One can legitimately claim that eagles have wings *in order to* fly; the reason is that their wings may have been maintained by natural selection because their effect (flying) may make a positive contribution to the survival and reproduction of their possessors. In this sense, it is legitimate to say that wings exist for flying, and, in this sense, teleology is perfectly legitimate in biology. What is the problem, then? The problem is not that the wings of eagles exist for a role (teleology), but how this role came to be. Insofar as one invokes natural processes, such as natural selection, to explain the presence of wings, there is no problem. The problem arises when one explains the presence of wings in terms of design: that an intentional agent made the wings exist for the purpose of flying. Organism parts, such as wings, can be said to exist *for* a purpose or role (teleology), but this is not the outcome of design.

Given these issues, conceptual change can be achieved if people are brought to conceptual conflict situations, where they could consider their intuitive explanations against the available evidence. Conceptual change is possible when people come to realize that there is a conflict between their own explanations and the available evidence, and that other explanations fit better with this evidence. For instance, conceptual change in evolution could take the form of a change in the concept of adaptation. The concept of adaptation that children and adults intuitively form is one that assumes intentional design. Thus, conceptual change in evolution could be the change in the concept of adaptation from being the outcome of intentional design to being the outcome of evolutionary processes, such as natural selection. To show how conceptual change in evolution can take place, let us consider another example: the shapes and the way of breathing of dolphins and sharks. By referring to dolphins and sharks, and not to "the dolphin" or to "the shark," the essentialist obstacle is addressed: Reference is made to a population (i.e., dolphins – plural) and not to individuals (i.e., dolphin – singular). Having done this, the focus is then on teleology.

If one asked why dolphins and sharks have hydrodynamic shapes, an intuitive answer would be *in order to swim fast underwater*. Dolphins and sharks are relatively large marine organisms that manage to swim fast underwater thanks to their hydrodynamic shapes. So, one could intuitively think that it is no coincidence that both dolphins and sharks have hydrodynamic shapes. They have similar sizes, are predators, and face similar difficulties and challenges in the underwater environment in which they live. They can survive if they can catch their prey and overcome competition, so swimming fast is one way to achieve this. Therefore, their hydrodynamic shape contributes to swimming fast, which in turn contributes to catching prey. In this sense, hydrodynamic shape can be said to exist in order to swim fast underwater. This can be generalized as follows:

> [T] Organisms *O* have trait *A* in order to perform function *B*. (T stands for Teleology)

As already explained in detail above, this is a legitimate proposition, and in fact it can also be a legitimate explanation for the existence of a particular feature. The important question is what underlies such a proposition. If such a proposition stems from the design stance, then the design-based explanation (design teleology, or DT) would have the general form:

> [DT] Organisms *O* have trait *A* in order to perform function *B*, because organisms have the features that are necessary for their survival.

Whereas the selection-based explanation (selection teleology, or ST) would have the general form:

> [ST] Organisms *O* have trait *A* in order to perform function *B*, because the latter confers an advantage; consequently, *A* has been selected for doing this and has been maintained in their lineage.

If we apply explanations [DT] and [ST] to explain the shapes and the way of breathing of dolphins and sharks, we would have the explanations presented in Table 3.1.

Question	Design teleology	Selection teleology
1. Why do dolphins have hydrodynamic shapes?	[DT1] Dolphins have hydrodynamic shapes in order to swim fast underwater, because organisms have the features that are necessary for their survival.	[ST1] Dolphins have hydrodynamic shapes in order to swim fast underwater, because the latter confers an advantage; consequently, this feature has been selected for doing this and has been maintained in their lineage.
2. Why do sharks have hydrodynamic shapes?	[DT2] Sharks have hydrodynamic shapes in order to swim fast underwater, because organisms have the features that are necessary for their survival.	[ST2] Sharks have hydrodynamic shapes in order to swim fast underwater, because the latter confers an advantage; consequently, this feature has been selected for doing this and has been maintained in their lineage.
3. Why don't dolphins have gills?	[DT3] Dolphins do not have gills, but have lungs in order to get more oxygen directly from the atmosphere, because organisms have the features that are necessary for their survival.	[ST3] Dolphins do not have gills because this feature was not maintained in their lineage and because lungs evolved in their terrestrial ancestors.
4. Why do sharks have gills?	[DT4] Sharks have gills in order to breathe underwater, because organisms have the features that are necessary for their survival.	[ST4] Sharks have gills in order to breathe underwater, because the latter confers an advantage; consequently, this feature has been selected for doing this and has been maintained in their lineage.

Table 3.1 DT and ST explanations for the features of sharks and dolphins

Obviously, proposition DT1 is compatible with DT2 and ST1 is compatible with ST2. However, propositions DT3 and DT4 are incompatible. Why would two organisms, which both live underwater, have different organs for breathing had they been designed (or, more generally, were created so as to have whatever feature is necessary for their survival)? On the other hand, propositions ST3 and ST4 are compatible with each other. So, when the explanatory scheme ST is used, it produces propositions ST1 to ST4, which are all compatible with one another. In contrast, when the explanatory scheme DT is used, some of the propositions produced (DT3 and DT4) are logically incompatible. Therefore, the design stance is simply explanatorily insufficient.

A simple way to illustrate this takes the form of the following narrative (which I once watched during a documentary film): A big gray whale was swimming in the ocean, close to the surface, with its newborn that was barely the size of a big dolphin. The newborn was swimming very close to its mother's body. If you ask anyone why these animals have hydrodynamic shapes, the most likely response will be that they have them in order to swim fast underwater. So far so good. Then, suddenly, two orcas, which are also mammals like the whales, approached the mother whale and the newborn and tried to separate them. The orcas did not get very close to the mother whale as it could hit them hard, and so tried for a long time to separate her and the newborn. Eventually they succeeded, and then repeatedly pushed the newborn deep into the sea until it drowned. But this would not have happened if gray whales had gills. The question that one can ask, then, is why don't whales have gills? The answer is simply that there are indeed some features that exist in order to perform a role because natural selection has favored the survival and reproduction of their bearers. However, organisms do not have all the features that they need in order to live in a particular environment. This is why dolphins and sharks, compared above, differ significantly in many characteristics, even though they live in similar environments. Dolphins have forelimbs, whereas sharks have fins; dolphins have mammary glands whereas sharks do not; dolphins have lungs whereas sharks have gills; dolphins have blowholes whereas sharks do not; and many more. Why would two kinds of organisms that live in the same environment be so different from each other? The answer is simple: Because they have come to be as they are through evolution, and were not intelligently and intentionally designed.

Making the shift from preconceptions to scientifically legitimate conceptions is far from simple. Many obstacles can obscure our understanding of life. It is therefore very interesting to see how Charles Darwin himself underwent a conceptual change process from his own views about purpose and design in nature to the theory of natural selection.

4 Charles Darwin's Conceptual Change

The Development of Darwin's Theory

Many people have heard of Charles Darwin (Figure 4.1) and his book *On the Origin of Species* (hereafter *Origin*). But I am not very confident that all those people who have something to say about Darwin's book (both proponents and opponents) have actually read it. The *Origin* was written by Darwin as an abstract of his species theory for a general audience. It is written with clarity, and it is full of insight and evidence for evolution. Reading the book also provides one with reflections of the particular political, cultural, social, religious, and scientific contexts in which Darwin's theory was developed and published. Darwin was a man of science; his aim was to convince his readers about natural selection as the process of transmutation (this is how the emergence of a species from a preexisting one was called at the time, whereas the word "evolution" referred to progress and development).

Darwin initially admired Paley's views, already discussed in Chapter 2. However, it seems that by March 1837 Darwin had become a convinced transmutationist, and in July 1837 he started his notebooks on transmutation, even though it took him 20 more years to arrive at the theory that we read in the *Origin*. This was accomplished through his extensive, careful, and insightful study of nature, as well as his wide reading and reflection on that reading. What was crucial for the development of his theory was his reading of and experience with breeding and artificial selection. Darwin explicitly referred to artificial selection in his definition of natural selection:

> I have called this principle, by which each slight variation, if useful, is preserved, by the term of Natural Selection, in order to mark its relation to

Figure 4.1 Charles Darwin.

> man's power of selection. We have seen that man by selection can certainly produce great results, and can adapt organic beings to his own uses, through the accumulation of slight but useful variations, given to him by the hand of Nature. But Natural Selection, as we shall hereafter see, is a power incessantly ready for action, and is as immeasurably superior to man's feeble efforts, as the works of Nature are to those of Art.

Darwin was not the first to consider animal and plant breeding as a source for understanding natural history. However, he was one of few among his contemporary men of science who was quite familiar with the work of breeders, although information on plant and animal breeding was widely disseminated in England at that time. Darwin managed to establish an extensive network of contacts that involved breeders, most of whom he never met in person. These people provided him with valuable information for his studies. At the same time, they benefited from Darwin who, as a leading man of science, lent them status by referring to their work in his writings.

Darwin first made use of the concept of selection in his Notebook C that covers the period from February to July 1838. After reading the pamphlets written by animal breeders John Sebright and John Wilkinson, who were explicit about the nature and power of artificial selection, Darwin realized that sustained selection for small changes could be taking place in nature. It was especially Sebright who mentioned natural selection, although by another name, and discussed the analogy between that and artificial selection. Darwin joined several pigeon-breeding clubs to see for himself how far selective breeding could go in producing new varieties. He thus realized that artificial selection could provide important insights about the process of transmutation. Darwin's analogy between artificial and natural selection was based on the assumption that breeders' selection resulted in modifications in the domesticated organisms that were permanent and that had not existed in their wild ancestors.

Eventually, in September 1838, Darwin read Malthus' *Essay on the Principle of Population* and came up with the idea of natural selection, which he developed in full by March 1839. Malthus argued that while the natural tendency of human populations was to increase in numbers at a geometric rate, agricultural production increased at an arithmetic rate. Consequently, there would be a struggle for resources that slowed population growth and hence limited the increase of human population. Although Darwin wrote in his autobiography that he happened to read Malthus for amusement, it seems that he had earlier been quite familiar with his views as there is considerable discussion of Malthus in Paley's *Natural Theology*, and as he had come to know people like Harriet Martineau who were promoting Malthusian ideas. Nevertheless, it was in September 1838 that Darwin had the insight about a struggle for existence taking place in nature that might result in selection similar to that employed by breeders. Eventually he wrote in the *Origin* that:

> Owing to this struggle for life, any variation, however slight and from whatever cause proceeding, if it be in any degree profitable to an individual of any species, in its infinitely complex relations to other organic beings and to external nature, will tend to the preservation of that individual, and will generally be inherited by its offspring. The offspring, also, will thus have a better chance of surviving, for, of the many individuals of any species which are periodically born, but a small number can survive.

The idea of struggle for existence thus became for Darwin the driving force for natural selection. However, there seems to be a significant difference between the ways in which he and Malthus conceived of this. Malthus' view of struggle was that of a species against its environment. Darwin rather conceived of two distinct types of struggle: The struggle of a species as a whole against its environment, and the struggle that resulted from the competition between individuals of the same species. Darwin's theory was based on a combination of these two types of struggle. The important insight that Darwin added was that the struggle between individuals of the same species was a consequence of the struggle of species against their environments.

This point was crucial for his analogy between artificial and natural selection, because these two processes differ in a very important aspect. Artificial selection requires an intelligent external selector who picks variants according to particular aims or goals. No such selector exists in the process of natural selection, which is the outcome of an unintentional, natural process of struggle among individuals. Therefore, the analogy between artificial and natural selection is weak. But according to Darwin, the competition between individuals of the same species that takes place simultaneously with competition between individuals of different species results in divergence. In this view, it is the external environment, and the different types of competition that it entails, that takes the place of the intelligent selector of artificial selection and that makes natural selection strongly analogous to artificial selection. Individuals of the same species interact with one another but also with other individuals from different species. In the long run, those individuals of a species that can compete more effectively in their environment with individuals of their own or of other species will be those that will live and reproduce and that will eventually be "naturally selected."

Darwin's reading of Malthus was important for the development of his theory for another reason. Two major representatives of the philosophy of science of the time were John Herschel and William Whewell. Both of them insisted that genuine science required an extensive evidential basis and the identification of a cause that could explain phenomena in different areas. They believed that the aim of science was to find the laws of nature and then to identify the true causes (*verae causae*) that guided the workings of these laws. What Darwin found after reading Malthus was quantitative laws, the best kind of laws according to Herschel and Whewell, leading to the idea of the struggle for existence on which Darwin could then base the analogy between artificial selection and natural selection. This was crucial for establishing the competence of natural selection to produce new species. However, he never managed to show that natural selection was actually responsible for producing new species through controlled observation and experiment.

Even though Darwin's initial harmonious view of nature changed after reading Malthus, he did not altogether reject the idea of harmony. Rather, he just abandoned the idea of perfect adaptation (that there is only one best possible form for any given set of conditions) for a similar view that nevertheless allowed the possibility of alternative forms and rudimentary organs (described as the idea of limited perfection). These views changed later when Darwin considered the developmental concepts and the generalizations of Karl Ernst von Baer and Henri Milne-Edwards. Von Baer had suggested that animals developed by a progression from a common pattern to a more specialized one. Milne-Edwards had argued that the diverging paths of development suggested by von Baer corresponded to the branching series of organisms in natural classification. Therefore, the process of development revealed natural relations that would best be represented with a branching arrangement. After considering these ideas, Darwin came up with the idea of the principle of divergence that incorporated the views of von Baer and Milne-Edwards, and at the same time was compatible with the idea of natural selection.

The core concept of the principle of divergence was the ecological division of labor, formulated between November 1854 and January 1855. By that time, Darwin had started organizing data from various sources and then drew conclusions and established hypotheses to be tested against these data. It seems that it was from these processes that the idea of the principle of divergence originated. This principle was an important innovation as Darwin needed to explain how natural selection could give rise to the various branches of the tree of life. He wrote:

> Here, then, we see in man's productions the action of what may be called the principle of divergence, causing differences, at first barely appreciable, steadily to increase, and the breeds to diverge in character both from each other and from their common parent. But how, it may be asked, can any analogous principle apply in nature? I believe it can and does apply most efficiently, from the simple circumstance that the more diversified the descendants from any one species become in structure, constitution, and habits, by so much will they be better enabled to seize on many and widely diversified places in the polity of nature, and so be enabled to increase in numbers.

In short, according to the principle of divergence, natural selection could indefinitely produce better-adapted forms by increasing the ecological specialization within groups, the members of which would eventually diverge from the initial form. Thus, natural selection would automatically increase the ecological division of labor among animals found in competitive situations, by favoring those individuals that were most able to exploit new niches. Interestingly enough, although the idea of the division of labor was used by the political economists of that time, such as Adam Smith, Darwin cited the zoologist Milne-Edwards instead, perhaps in an attempt to provide scientific foundations for his theory. However, it has been suggested that Darwin's use of this term is closer to that of Adam Smith than to that of Milne-Edwards. This idea is depicted in Figure 4.2, as well in the branching diagram that is the only figure included in the *Origin*.

The principles of natural selection and divergence are the central ones in Darwin's theory, but he came up with them more than 15 years apart. To understand why this was the case, we need to consider the conceptual shifts he underwent during that time.

From Natural Theology to Natural Selection

Darwin had no "Eureka!" moment. The development of his theory was a very long process. Although he started taking notes on transmutation in July 1837, and came up with the idea of natural selection by March 1839, it took him until November 1854 to come up with the wider (macroevolutionary) pattern

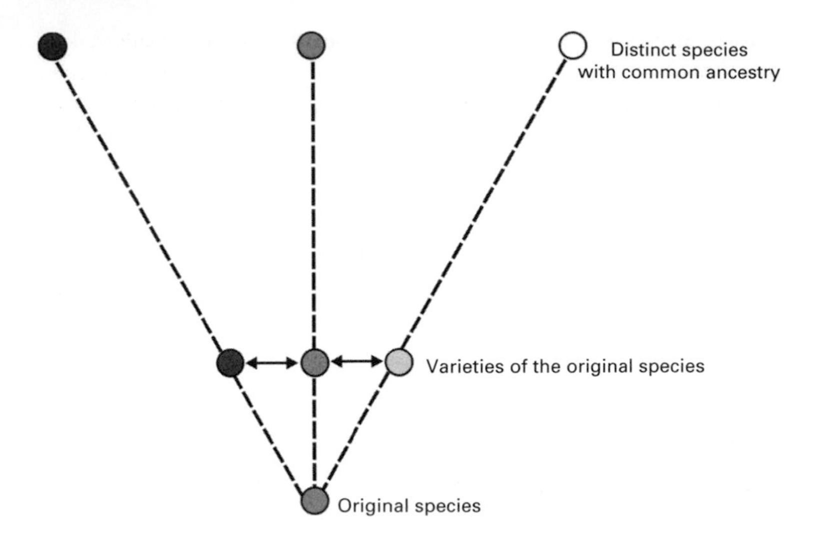

Figure 4.2 How varieties gradually diverge and become distinct species.

in which natural selection would fit. Overall, it seems that there are two important shifts in Darwin's views: (1) the shift from special creation to transmutation, and a while later to natural selection; and (2) the shift from perfect adaptation to relative adaptation. One might assume that the first shift was the most important one. After all, it is then that Darwin became an evolutionist. However, it is one thing to conceive of natural selection, and another to be able to provide a coherent and well-grounded account of how it takes place. The latter depended on the second shift mentioned above. Let us then consider in some detail how these conceptual shifts occurred.

A significant piece of evidence for Darwin's theory came from the geographical distribution of species. Perhaps the most well-known case, and one that actually had a major influence on Darwin's changing views about the origin of species, was that of the Galápagos Islands. His views did not change when Darwin visited the islands in September and October 1835, but after his return to England the next year. Darwin was wondering whether the birds he had collected from these islands were distinct varieties or distinct species. Concluding that they were distinct varieties would have important implications for the possibility of transmutation, because varieties have the potential to give rise to new species. If this was not the case and the birds were distinct species, he would assume that they could have simply migrated there from the American continent. However, in March 1837 ornithologist John Gould told Darwin that the birds he had collected at the Galápagos Islands were distinct species that were not found on the American continent. Therefore, these species could have only originated on the Galápagos Islands. In addition, the bird species that had originated on those volcanic islands were very similar to species that had already originated under very different conditions on the nearest older continental land, rather than to species that had originated on other volcanic oceanic islands elsewhere in the world. For Darwin, such similarities could only be explained as the outcome of common ancestry and not as adaptations to common conditions. This was crucial evidence against the special creation of species.

The Galápagos Islands actually provided several examples against this idea. Why did those islands, which are far from a mainland, lack certain kinds of organisms, such as amphibians, although these could certainly live under the conditions there? It is probably not a coincidence that amphibian adults and eggs are killed in salt water and so they could not be accidentally transported across the ocean. Similarly, why are these remote islands characterized by a high proportion of endemic species – that is, species that are found only there and not in other places in the world – compared to those close to a mainland? Most interestingly, why are these species always closely related to those found on the nearest mainland, in the case of the Galápagos similar to those found on the American continent, although the conditions on the islands are very different from those on the mainland? According to Darwin, island populations emerged as a result of migration from the nearest mainland, and this explained the similarities with the organisms living there. But since migration did not take place all the time, island populations also diverged from those living on the mainland and eventually became distinct species. In addition, despite the similar physical conditions on the islands, different organisms might migrate to different islands at different times and find different competitors or food sources. Thus, the initial populations could evolve in different directions, and because organisms only rarely moved between islands, populations could eventually diverge and become distinct species. Overall, the observations at the Galápagos Islands were not consistent with the idea that species were specially created to be perfectly adapted to their environments.

As already mentioned, Darwin initially accepted the idea of perfect adaptation, according to which there was one best possible form for any given set of conditions. The first shift in his views occurred after his reading of Malthus, near the end of September 1838. However, this was still a view of perfect adaptation, but one in which perfection was limited. After reading Malthus, Darwin wrote that the final cause of the process described "must be to sort out proper structure, & adapt it to changes." And these "proper structures" are forced into the gaps in the economy of nature, while new gaps are formed when other structures die out. This is an argument based on design: Adaptation is the final cause of the struggle for existence. And this adaptation is perfect as it is not determined by the environment. It is already formed, sorted out, and "forced" into the environment. Thus, at that time Darwin thought of natural selection as guiding form toward perfection. He assumed that natural selection made species perfect for the place they occupied, as well as that there was no variation among perfectly adapted forms. Natural selection was thus compatible with the idea of a harmonious nature as long as it was perceived as a law of nature that was aimed to produce perfect adaptation. This idea retained the notion of harmony, but allowed the possibility of alternative forms and rudimentary organs. But as soon as Darwin came up with the principle of divergence, he realized that adaptation must be relative. Not only was the adaptedness of a species relative to the adaptedness of other species living in the same area, but also a well-adapted species could eventually become even better adapted through natural selection. The idea of relative adaptation was a necessary implication of the principle of divergence: New individuals that could exploit new niches were favored by selection, and so there was divergence of form. But in this case, the direction of divergence of form was relative to the environment, as only those who could adapt to it would survive. This second shift seems to be evident for the first time in his (never completed) book *Natural Selection* around March 1857.

This is related to another important shift in Darwin's views about the amount of variation available in nature. In 1844 Darwin thought that there was little available variation, as perfectly adapted forms did not vary. Consequently, a change in external conditions was required for new variation to occur. In 1859 he assumed that no change was required and that variation was common. A consequence of this, and a third difference between 1844 and 1859, was that Darwin initially thought that transmutation could take place only in response to changes in external conditions (either because the environment itself changed or because organisms migrated to a new environment). In contrast, in 1859 Darwin believed that since variation was common, natural selection could take place at any time, leading to transmutation and to the production of better-adapted forms. A major reason for this second shift may have been Darwin's extensive study of barnacles from October 1846 to September 1854. This study made Darwin realize that there was more variation available in nature than he initially thought. Given this, not only did the idea of perfect adaptation seemed to be less plausible, but there was also adequate variation for natural selection to occur.

Finally, it has also been suggested that a reason for the shift from his initial natural theological assumptions may have been the death of his beloved daughter Annie in 1851, who died at the age of 10 after having probably suffered from tuberculosis. Although this tragic event caused Darwin terrible pain, he did not seem to experience any instant loss of faith. Instead, it seems that he rather went through fluctuations of belief that were also influenced by his social circle, which included people who could lead a moral life without embracing Christianity, and of course his own understanding of natural selection producing suffering that was inconsistent with a benevolent God. Whether Darwin's pain from the tragic loss of Annie made him abandon his belief in a harmonious world or not is a matter of speculation. There is no question that such a terrible loss might make one reconsider one's views about harmony and perfection in the world. However, it may be the case that this incident was not as influential as commonly thought. The major phases and points of shift in Darwin's thinking are presented in Figure 4.3.

The important question now is: What caused these conceptual shifts? It is Darwin's realization that his initial conceptions about design in nature were explanatorily insufficient. He considered this in his *Autobiography*:

> The old argument of design in nature, as given by Paley, which formerly seemed to me so conclusive, fails, now that the law of natural selection has been discovered. We can no longer argue that, for instance, the beautiful hinge of a bivalve shell must have been made by an intelligent being, like the hinge of a door by man. There seems to be

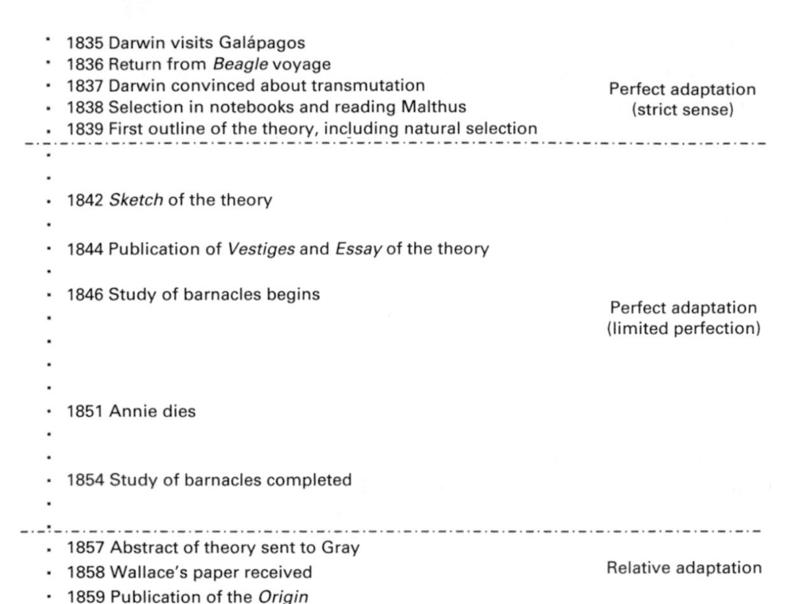

Figure 4.3 The major phases and shifting points in Darwin's thinking.

> no more design in the variability of organic beings and in the action of natural selection, than in the course which the wind blows. Everything in nature is the result of fixed laws.

Darwin's extensive study of nature showed him that species cannot be specially created, as well as that the world is not as harmonious as he initially believed. Whereas he initially believed that organisms are perfectly adapted, he came to realize that many organisms could be maladapted or not adapted at all. This made him change his initial views. He accepted transmutation in the place of special creation and he developed a theory of descent with modification based on natural selection. He also incorporated imperfections in the idea of perfect adaptation, which was eventually replaced by the idea of relative adaptation. His conclusion was that populations were not adapted in any absolute sense, but only relatively to the environment they inhabited. And these populations became well adapted through natural selection.

Darwin's explanation for the origin of organisms' characteristics based on their evolutionary history was more efficient than design (remember the discussion in Chapter 3 about the hydrodynamic shape of dolphins and sharks and the comparison between propositions DT and ST). But he made no further conclusion that God does not exist. Indeed, he wrote in his autobiography: "I cannot pretend to throw the least light on such abstruse problems. The mystery of the beginning of all things is insoluble by us; and I for one must be content to remain an Agnostic." Darwin's conceptual change was thus one from a view of organisms as artifacts, intelligently designed for a purpose, to their view as natural entities; it was not a shift from theism to atheism.

The Publication of the *Origin of Species*

Even though Darwin had developed some core concepts of his theory by 1839, he did not publish it until 1859. Whereas there has been some speculation that this was due to his fear of the public reaction to his views, historians have shown that this delay was not just a matter of that. Darwin was certainly concerned about the reaction of religious people who might consider his theory – which entailed that humans were just one species closely related to primates and not one specially created by God – as an attack on the established beliefs of the time. One of these people was his wife, Emma Darwin (née Wedgwood), who was deeply religious, and who believed in resurrection and salvation. Because Darwin's scientific findings on the origin of humanity and Emma's own Christian beliefs were in conflict, she was afraid that his ideas would keep them apart in the afterlife. In 1844 Darwin gave Emma an essay containing an outline of his theory, which was an enlarged version of a sketch he had written in 1842. He also gave her a letter in which he asked her to publish this essay in the event of his sudden death. He seemed to prefer giving up credit for his ideas during his lifetime rather than hurting her feelings or being the cause of her and their children's social ostracism. Darwin also shared his views with Joseph Dalton Hooker, a botanist and future director of Kew Gardens, in a letter that he wrote in the same year, admitting that "I am almost convinced (quite contrary to the opinion I started with) that species are not (it is like confessing a murder) immutable."

But Darwin was also concerned about the reaction of the leading scientific figures of his day. He was well aware of the reaction to previously published theories of evolution. For instance, Jean-Baptiste Lamarck's theory had been severely criticized by Charles Lyell, a prominent geologist, in his *Principles of Geology*. Lyell had written a long and careful exposition of Lamarck's theory, and questioned its evidential basis. There was a fiercer reaction against the *Vestiges of the Natural History of Creation*, anonymously published by Robert Chambers in 1844. Thomas Henry Huxley, a proponent of evolution, was very critical in his 1854 review of a later edition of the *Vestiges*: "we find reason to doubt if the author ever performed an experiment or made an observation in any one branch of science." The public reaction to the *Vestiges* made Darwin uncomfortable about the prospect of publishing his own evolutionary views. He was aware that his book would be judged in comparison to the *Vestiges*, which for many was an attack on Christianity.

Most importantly, Darwin was aware that these older theories had been criticized because they were mostly speculative and were not based on solid scientific research. Therefore, he tried to gather sufficient data in support of his theory; his work on the classification of barnacles, which lasted for eight years, was in part a response to the reviews of the *Vestiges*. His study of the barnacles was crucial for a major scientific problem he had to resolve: that of limited variability and of subsequent weak natural selection. Darwin arrived at important conclusions from his barnacle research, such as that there was a high degree of variation in all external characteristics. The study of the barnacles also gave Darwin scientific credibility as he won the Royal Society's Royal Medal for Natural Science. Classification was seen as the foundation of natural history, and so by establishing his expertise in this field, Darwin also established his competence as a naturalist.

Therefore, it becomes clear that Darwin did not have a complete theory at hand in 1839, and this is why he refrained from having it published. Rather, he had questions to answer and problems to solve, which took some time and resulted in a theory quite different from the one he initially came up with. In 1856, he started working on a big book that would be called *Natural Selection*. However, this was interrupted by the receipt in 1858 of a letter from Alfred Russel Wallace. This was the incident that eventually forced Darwin to proceed to the publication of the *Origin*. Wallace was one of Darwin's numerous correspondents from around the world who knew that Darwin was interested in the question of how species originate, and trusted his opinion on the matter. Thus, Wallace sent Darwin his essay in which he presented his own answer to this problem and asked him to review it. While

Wallace's essay did not employ Darwin's term "natural selection," it did outline a process of evolutionary divergence of species from preexisting ones due to environmental pressures. In this sense, it seemed the same as Darwin's theory, although it was quite different in some crucial aspects.

Nevertheless, Darwin considered Wallace's theory as being the same as the theory he had worked on for 20 years but had yet to publish; he wrote in a letter to Charles Lyell: "if Wallace had my MS. sketch written out in 1842, he could not have made a better short abstract! Even his terms now stand as heads of my chapters!" Darwin's priority was eventually saved as Lyell and Hooker arranged for a joint presentation at the Linnean Society of both Darwin's and Wallace's papers. Wallace found out about this several months later, and sent Darwin a letter of approval that arrived early in 1859. Wallace was not at the presentation, but neither was Darwin because one of his children was seriously ill. It should be noted that Darwin's priority in conceiving natural selection was certified by an abstract of his theory that he had sent to the botanist Asa Gray as early as September 1857, which was part of what was presented to the Linnean Society. Eventually, the *Origin* came out on November 24, 1859.

As already described in the previous sections, the concepts of struggle for existence, artificial selection, and divergence form the conceptual foundations of the *Origin*. The argumentation in the *Origin* involved two central concepts: the tree of life and natural selection. According to the first concept, species changed over time; some went extinct while others continued to exist or gave rise to multiple descendant species. This concept involved two other distinct concepts: transmutation (one species changing into another) and common descent (one species splitting into two or more species). The second central concept, natural selection, explained how species changed through a process of selection similar to that applied by breeders to domesticated varieties of plants or animals. The arguments in the *Origin*, and the respective analogies on which they are based, are presented in Figure 4.4.

It is important to note that there were many intellectual and practical resources, characteristic of Victorian society, which were not available in earlier times and which were crucial for the development of his theory. Darwin's influences were distinctly of the Victorian era: Thomas Malthus and Adam Smith were political economists who developed their theories in

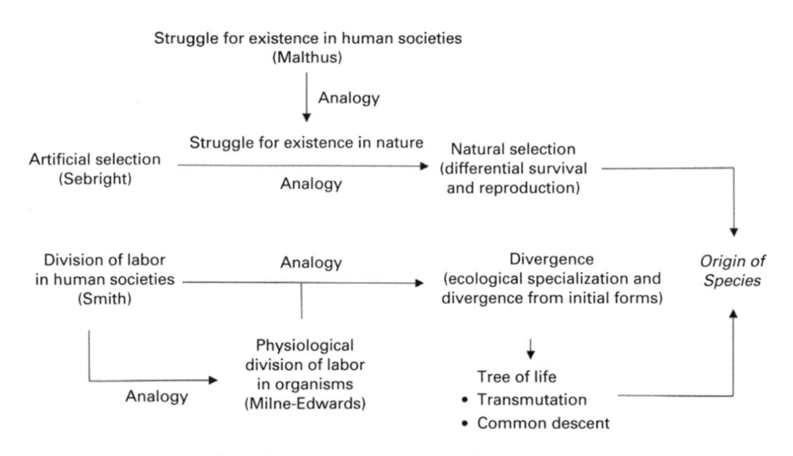

Figure 4.4 The conceptual foundations and the arguments in the *Origin*. Darwin's theory was based on particular analogies: (1) between the struggle for existence in human societies and the struggle for existence in nature; (2) between artificial selection and natural selection; and (3) between the (physiological) division of labor and ecological specialization.

the English context; animal and plant breeding was a form of Victorian technology; John Herschel and William Whewell were among the first philosophers of science in a tradition based on Newton's science; and the Anglican Church had a tradition – natural theology – that put emphasis on the idea of adaptation. Furthermore, industrialization and imperialism of the British Empire created many of the conditions that made Darwin's research possible. He famously traveled around the world for five years aboard HMS *Beagle*. He also developed a vast network of correspondents, most of whom he never met in person, thanks to the British ships traveling around the world. There was a public interest for natural history, which became very popular in Victorian England because it was seen both as amusement and as science. The expansion of the railway network facilitated access to various sites where specimens could be collected. Moreover, the Penny Post made the exchange of specimens possible between naturalists from various parts of the country. As the practice of natural history required not only considerable skill but also specialist equipment, technological advancement also took place. Cheaper books were soon produced, many of which were about natural history. Darwin made good use of all these resources. Given all this, it is no coincidence that Alfred Russel Wallace, who independently came up with a similar theory, lived at the same time and in the same culture as Darwin. But it was Darwin's background, knowledge, experience, and skills that brought him to a unique position to come up with the theory that we read in the *Origin*.

Science and Religion in the Reviews of the *Origin of Species*

It is often thought that the criticisms of the *Origin* were made strictly on religious grounds, because of the implications that the theory had for the status of humanity. However, this is not the whole story.

Let us consider the following two excerpts from reviews of the *Origin*:

> After much consideration, and with assuredly no bias against Mr. Darwin's views, it is our clear conviction that, as the evidence stands, it is not absolutely proven that a group of animals, having all the characters exhibited by species in Nature, has ever been originated by selection, whether artificial or natural. Groups having the morphological character of species distinct and permanent races in fact have been so produced over and over again; but there is no positive evidence, at present, that any group of animals has, by variation and selective breeding, given rise to another group which was, even in the least degree, infertile with the first. Mr. Darwin is perfectly aware of this weak point, and brings forward a multitude of ingenious and important arguments to diminish the force of the objection.

And here is the second one:

> Further, the embalmed records of 3000 years show that there has been no beginning of transmutation in the species of our most familiar domesticated animals; and beyond this, that in the countless tribes of animal life around us, down to its lowest and most variable species, no one has ever discovered a single instance of such transmutation being now in prospect; no new organ has ever been known to be developed – no new natural instinct to be formed – whilst, finally, in the vast museum of departed animal life which the strata of the earth imbed for our examination, whilst they contain far too complete a representation of

> the past to be set aside as a mere imperfect record, yet afford no one instance of any such change as having ever been in progress, or give us anywhere the missing links of the assumed chain, or the remains which would enable now existing variations, by gradual approximations, to shade off into unity.

These two excerpts actually make the same point: that there is no strong evidence that natural selection has actually brought about the emergence of new species in nature. Indeed, Darwin never managed to establish this. Wouldn't you agree that these two excerpts could come from the same review of the *Origin*, written by the same person(s)? However, these excerpts come from two different reviews written by two people who were entirely opposed to each other with regard to their ideas about evolution. The first one was written (anonymously) by Thomas Henry Huxley, one of the most prominent of Darwin's supporters, and the other was written (again, anonymously) by Samuel Wilberforce, Bishop of Oxford. Huxley acknowledged the importance of Darwin's contribution, but also pointed out that even if there was enough evidence for the competence of natural selection to produce new species, there was no actual evidence that it had indeed done so. In the same spirit with Huxley, Wilberforce questioned the competence of natural selection to produce new species, as there was no evidence that it has ever been actually responsible for doing so.

The legend has it that Wilberforce attempted to ridicule Darwin and his theory at a meeting of the British Association in Oxford on June 30, 1860. There, he faced Huxley, who is said to have succeeded in defeating Wilberforce and, through that, any attempt to impose on scientists the conclusions they were allowed to reach. This episode is widely cited as an instance of the conflict between science and religion. However, careful historical analysis has shown that the legend overlooks the fact that Wilberforce's speech, rather than reflecting prejudice and religious sentiment, included many of the scientific objections of Darwin's contemporaries. It also seems that Hooker's contribution in defending Darwin was more successful than Huxley's. Therefore, rather than being an instance of a wider conflict between science and religion, the Huxley–Wilberforce debate reflects trends and developments in Victorian society that had to do with the formation of science as a profession and particular divisions and reactions within the Church.

That the same criticism was made both by Huxley and Wilberforce in their reviews of the *Origin* is important for two reasons. First, it shows that scientific judgments and criticisms can be made on objective grounds by people with vast differences in their worldviews. Their motivations notwithstanding, both Huxley and Wilberforce raised a question that Darwin already knew would be raised: Has natural selection actually produced new species? Second, this shows that there was more in the reaction to Darwin's theory than religious instinct and fundamentalism. Not all religious people are ignorant and fundamentalist, and proponents of evolution must take their arguments seriously when they point to actual problems and difficulties of scientific theories.

Richard Owen, one of the leading scientific figures of the time, also wrote his own anonymous critical review of the *Origin*:

> If certain bounds to the variability of specific characters be a law in nature, we then can see why the successive progeny of the best antlered deer, proved to be best by wager of battle, should never have exceeded the specific limit assigned to such best possible antlers under that law of limitation. If unlimited variability by "natural selection" be a law, we ought to see some degree of its operation in the peculiarly favourable test-instance just quoted . . . We have searched in vain, from Demaillet to Darwin, for the evidence or the proof, that it is only necessary for one individual to vary, be it ever so little, in order to [sic] the conclusion that the variability is progressive and unlimited, so as, in the course of generations, to change the species, the genus, the order, or the class. We have no objection to this result of "natural selection" in the abstract; but we desire to have reason for our faith. What we do object to is, that science should be compromised through the assumption of its true character by mere hypotheses, the logical consequences of which are of such deep importance.

Owen questioned the competence of natural selection to produce new species, starting from variation in a single individual.

The above reviews support an important conclusion: that it is indeed possible to distinguish between what one knows and what one believes. Huxley described himself as an agnostic, and he was trying to turn science into a profession so that people other than clergymen could get employment in universities. His main aim was to liberate the practice of natural history from the domination of the Church. Wilberforce was a bishop, so there is no need to get into details about his religious beliefs. There were clergymen such as Charles Kingsley and Baden Powell who gladly accepted the idea of evolution; Wilberforce was for various reasons opposed to that. Owen was one of the greatest anatomists of his time, a prominent expert on fossils, the first superintendent of the natural history department of the British Museum, and also the person who convinced the government to found the London Natural History Museum. He made major contributions to science by revealing homologies that he ended up explaining as instances of the archetype in God's mind. And yet, as is shown by the above quotes, all of these people came to the same conclusion regarding Darwin's theory and questioned the competence of natural selection in producing new species, as it had never been shown to be responsible for doing so.

In my view, these quotes are good examples of how people should talk about science. Their (probably religious) motivation notwithstanding, Wilberforce and Owen raised important questions that Darwin did not manage to answer. His explanation for the origin of species was questioned on rational grounds. Natural selection seemed to be competent to produce new species, and thus was a plausible explanation for some observations (e.g., biogeography), but it had never been shown that it was actually responsible for doing so. Thus, natural selection could not be a true cause according to Herschel and Whewell. In the same spirit, no scientist today would accept an explanation as definitive if it was never shown that what is presented as a cause does indeed have particular effects. But since Darwin's time, numerous examples of evolution in general, and natural selection in particular, in action have been described.

Here is, then, what people should do: Examine the evidence available, decide whether a theory has a solid evidential basis, and then wonder about its implications beyond the realm of science. Evolutionary theory does have implications for personal worldviews, but one must first understand the scientific evidence before considering its broader implications. The take-home message, then, is that one should try to distinguish between knowledge and belief, or between scientific arguments and religious sentiment. And as the foregoing quotations show, it is possible to criticize a theory on scientific grounds independently of one's beliefs. On the one hand, even as ardent a Darwinian as Huxley did not blindly accept all of Darwin's propositions. On the other hand, religious believers like Wilberforce and Owen criticized the *Origin* solely on scientific grounds. It is therefore possible for scholars to debate on scientific–epistemic grounds despite their (sometimes entirely contrasting) religious beliefs, and this is what we should all try to do.

Let us now examine some core concepts of evolutionary theory, beginning with common ancestry.

5 Common Ancestry

The Evolutionary Network of Life

Evolutionary theory suggests that all species, both extinct and extant ones, have evolved through natural processes and are more or less related. There must have been one or a few universal common ancestor(s) of all species because organisms share some crucial characteristics: (1) they consist of (one or more) cells, or depend on cells for their reproduction; (2) they exhibit the characteristic properties of life (metabolism, reproduction, homeostasis, etc.), which are the outcome of intra- and/or intercellular processes; (3) proteins have central roles in these processes; and (4) these proteins are synthesized inside cells on the basis of specific DNA sequences and their interaction with their cellular contexts. These common characteristics are fundamental to life on Earth, but how could they have emerged? Creationists would argue that they are products of design; these were characteristics of an archetypal plan, on the basis of which living organisms were formed. Evolutionists would reply that no competent designer would design something with fundamental problems or imperfections (remember the comparison between dolphins and sharks in Chapter 3). But couldn't it be the case that the aforementioned characteristics (1–4), which are common in all organisms and fundamental to life on Earth, are the products of a designer who created the molecular and cellular foundations of life and then let it evolve?

Most likely, no. There is so much diversity in the molecular and cellular levels that is better explained as the outcome of evolution rather than of intentional design. For example, at least 5 percent of the human genome consists of families of DNA sequences which are more than 90 percent identical to each other. The DNA sequences implicated in the synthesis of rRNA are present in more than 400 copies in the human genome, and so an adequate amount of rRNA and eventually of ribosomes, where such a fundamental process as protein synthesis takes place, is produced. Also, several DNA sequences are involved in the production of the subunits of oxygen-carrying proteins (hemoglobin A, hemoglobin A_2, hemoglobin F, myoglobin), which are specialized for particular tissues or particular stages of development. It might at first seem that having multiple copies of similar DNA sequences is an advantage. However, there are cases where multiple copies of similar DNA sequences are responsible for a number of disorders. Because these DNA sequences are very similar to each other and because they are also arranged close to each other, they cause the abnormal pairing of the respective chromosomes during cell divisions. As a result, various changes in the structures of chromosomes (deletions, additions, inversions, or translocations of DNA sequences) occur, which are associated with a number of disorders such as α-thalassemia, hemophilia A, neurofibromatosis type 1, red–green color blindness, Prader–Willi syndrome, and others.

The presence of multiple copies of DNA sequences in our genomes is better explained as the outcome of evolution (e.g., unequal crossing-over of chromosomes has produced extra copies of existing sequences, which later accumulated changes and might have acquired new roles or not). In fact, all similarities between organisms at the molecular level are best explained through evolution from common ancestors. You can think of the genomes of all organisms as books that contain text written with the same letters (DNA nucleotides), which are combined to produce more or less the same words (almost universal genetic code: the same nucleotide triplets correspond to the same amino acid in most cases). Therefore, the different texts are quite similar not only because they contain the same words, but also because these are used to form quite similar sentences (DNA and protein sequences).

Let me elaborate on this argument. Consider this: The English and French alphabets are identical in terms of the letters they include, although they are pronounced differently. We know of course that these two alphabets are derived from Latin and their similarities are thus not coincidental – they are due to common descent. Nevertheless, these two identical alphabets give rise to very different words for the same concepts. Think of the numbers 1, 2, 3, and 4: The words that correspond to these numbers in English are "one," "two," "three," and "four," respectively, whereas the respective words in

French are "un," "deux," "trois," and "quatre." Not identical for sure. Despite the fact that some words are spelled identically in these two languages (although they may be pronounced differently, e.g., the word "table" or the word "impossible"), most words that correspond to the same concept are different. In most cases, the same message is transmitted through very different words that are nonetheless practically based on the same alphabet. Now, if two closely related languages can be so different in terms of words and sentences even though they are based on the same alphabet, the only rational conclusion is that the very similar DNA language used in all forms of life, which is framed using exactly the same alphabet (A, T, C, G) and sometimes almost the exact same words (DNA, RNA, and protein sequences), is evidence for their close relatedness and therefore their common ancestry.

Let us consider another example. Imagine two families, one coming from Europe and the other coming from Africa. Imagine also that each family consists of four members: a father, a mother, a daughter, and a son. The European parents are white with brown, straight hair. Their children are neither identical to their parents nor to each other, but they exhibit the aforementioned characteristics (white skin and brown, straight hair). In contrast, the African parents have brown skin and black, curly hair. Again, their children are neither identical to their parents nor to each other, but they exhibit the aforementioned characteristics (brown skin and black, curly hair). Why do children resemble their parents? Genetics provides the answer: Offspring develop from a fertilized ovum, a single cell that emerges from the fusion of the reproductive cells of their parents. Consequently, the genetic material of that first cell consists of the DNA molecules contained in the spermatozoon and the ovum of the parents, and each offspring possesses a unique combination of approximately one-half of the maternal DNA and one-half of the paternal DNA. Eventually, during development a multicellular organism emerges and particular parts of this DNA interact with their cellular environment to drive the formation of tissues and organs. Offspring resemble their parents in some respect because they have inherited part of their genetic material, but they will also be different from either or both of them due to the specific interactions of paternal and maternal DNA molecules, as well as due to several important phenomena taking place during development.

Family members are usually depicted in family trees. Parents are connected to each other with a horizontal line, whereas both of them are connected to their children with vertical lines. Males are depicted with rectangles and females with circles. Family trees are useful because one can infer shared characteristics based on relationships, as well as infer relationships based on shared characteristics. It is the latter that is of utmost importance for evolutionary biology. Scientists do not know the exact relationships between (extant or extinct) organisms and so they rely on common characteristics to make inferences about relationships. For instance, if you saw a family tree with all members having the same skin color and hair type, and another with both parents having a different skin color and hair type from both of their children, could you infer which one is more accurate? You would probably infer that the former is more likely to be accurate. It is not impossible of course for children to differ in color from their parents, but it is more probable for offspring to resemble their parents in terms of skin color, which is an inherited characteristic. *The first important conclusion is, therefore, that relatedness can be inferred from shared characteristics*.

It should be noted at this point that the grouping performed here was based on a characteristic chosen arbitrarily: skin color and hair type. If we decided to use another inherited characteristic, such as human blood groups, the eventual grouping could be different or we might end up with different kinds of groupings. For instance, let's assume that we decided to group these individuals on the basis of their blood groups. Let's assume that the African individuals all had blood group A and the European individuals all had blood group O; as well as that, we know nothing about the skin color of these individuals. Could we then conclude that all individuals with blood group A are members of the same family and that all individuals with blood group O are members of the same family as well? The answer is no. Although parents with blood group O can only give birth to children with blood group O, it may be the case that parents with blood group A give birth to a child with blood group O. Consequently, either of the European children could, in terms of grouping based on blood groups only, belong to the African family. *The second important conclusion, then, is that we cannot use any characteristic for grouping; some characteristics may be more appropriate than others.*

Their difference in skin color notwithstanding, could these families, and consequently their family members, be somehow related? Of course. All members of these families are humans; they have two eyes, two ears, a nose, hair on the top of their heads, more than 99 percent of their DNA sequences in common, and so much more. Consequently, each member of the European family is related to each member of the African family. And the difference in relatedness between any two individuals is a matter of degree, not of kind. Both the European and the African children belong to a wider group – humans – because, except for their parents who are their most recent ancestors, they also have some more remote common ancestors from whom these common characteristics are derived – the first humans. This makes the two families related as well. Each child is more related to the members of his/her own family and less related (but related nevertheless) to the members of the other family. Of course, in this way they are also related to every other human being in the world. In this sense, all organisms living on Earth are related; it is just that some are more closely related than others.

In order to depict the evolution of life on Earth and the relatedness of the various taxonomic groups (taxa) scientists use a different means of representation that has some crucial similarities with family trees: evolutionary trees. There are different kinds of evolutionary trees, such as phylogenetic trees and cladograms. Phylogenetic trees depict actual evolutionary histories, and their branches have different lengths that show the relative age of extinction. The branches of cladograms constitute hypotheses of relative recency of common ancestry, and ancestral or extinct taxa are located at the tips of terminal branches like extant taxa. Most of the trees used in this book are cladograms, and they are described as evolutionary trees. What is important to note is that the branching points (nodes) in evolutionary trees correspond to common ancestors, and that evolutionary trees indicate historical relationships, and not just similarities. Closely related species generally tend to be similar to one another; however, this is not always the case. For instance, crocodiles look more similar to lizards, but they are more closely related to birds when DNA sequences are used for the comparison. In this case, relatedness refers to common ancestry: The more recent two species' common ancestor, the more closely related they are.

Family trees share crucial similarities with evolutionary trees, despite the major difference that the common ancestor in an evolutionary tree is a single taxon (taxonomic group), such as species or genus, whereas the common ancestor in a family tree is a couple. Because family trees are familiar to all of us, I am using them here to help readers understand the less familiar evolutionary trees. The first similarity is that family trees indicate relatedness on the basis of common ancestry. The European girl is more closely related to her brother than to the African girl because the (most recent) common ancestors she shares with her brother, their parents, are more recent than the (more remote) common ancestor she shares with the African girl. Thus, both family trees and evolutionary trees provide historical information so that one can infer relatedness from how old the common ancestor is. Another similarity is that both family and evolutionary trees facilitate grouping. In family trees we can group individuals into families; a family could consist of a couple and their children, or of a couple and their children, grandchildren, and great-grandchildren, or of an individual and his/her parents, his/her grandparents, and his/her great-grandparents, etc. In evolutionary trees, taxa can be grouped in clades, which are hierarchically nested groups that include a common ancestor and all its descendants. To illustrate this in terms of a family tree, a group that includes a couple, all their children, and all their grandchildren would be a clade; if the group included a couple, one of their children, and his/her offspring but not the other one and its offspring, then this would not be a clade. Therefore, all members within a clade share at least one common (usually identifying) characteristic that is derived from the common ancestor.

Let us now see how evolutionary trees are formed, by way of analogy with family trees. To do this, some assumptions are required: We can think of clades as families, and of species as individual members of families; births will correspond to speciation events, i.e., to the production of new species, and both parents will correspond to the common ancestor. Thus, let's imagine a couple (I1 and I2) that gives birth to two children (II2 and II3), each of which also gives birth to two children (III1, III2, III3, III4 – to illustrate the differences, the offspring of daughter II2 are two girls whereas the offspring of son II3 are two boys) (Figure 5.1a). This family tree can, under the above assumptions, be represented as an evolutionary tree (Figure 5.1b – circles are used for both males and females here) that is equivalent to an actual one (Figure 5.1c), which in turn can also be depicted in a branching form (Figure 5.1d). Grandparents (I1 and I2) correspond to the earlier common ancestor (G) of all species A–D; parents (II2 and II3) correspond to the most recent common ancestor P of species A and B, and to the most recent common ancestor Q of species C and D, respectively. Finally, children (III1, III2, III3, III4) correspond to species A–D. Species are genetically more similar to their more recent common ancestor than to the older one, as children are genetically more

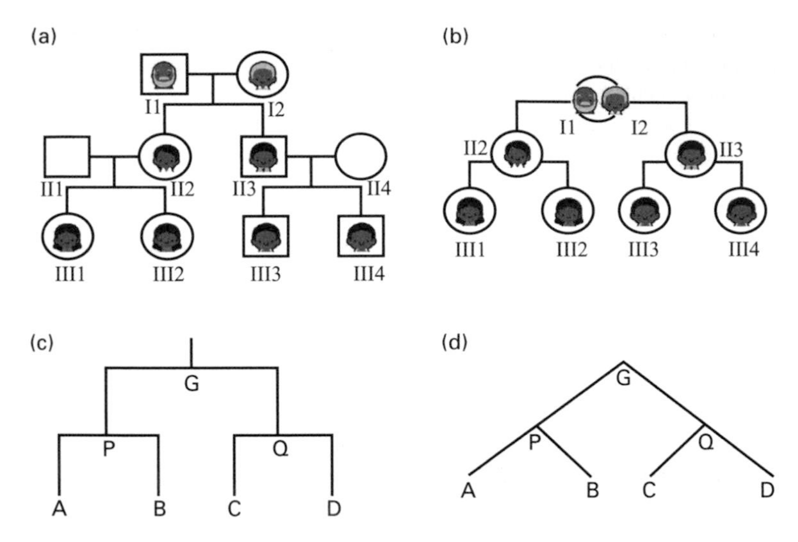

Figure 5.1 Analogy between family trees and evolutionary trees. A family tree (a) is gradually (b) transformed to an evolutionary tree (c) in order to show the similarity in the process of reconstructing human and species ancestry (under some abstract assumptions, of course – see text for details).

similar to their parents than to their grandparents. Species A and B, as well as species C and D, are not identical to each other as two brothers or two sisters are not identical to each other (except for identical twins, which is not the case here).

As I have already mentioned, a central assumption of evolutionary theory is that all life has a common ancestry. We currently have a good idea of what this common ancestor could be. Organisms are usually divided into two major groups, based on the morphological characteristics of their cells: prokaryotic (cells without nuclei; this group practically includes bacteria only) and eukaryotic (cells with nuclei – this group practically includes everything except for bacteria). Organisms are nowadays classified in three domains: Archaea, Bacteria, and Eukarya (or archaeobacteria, bacteria, and eukaryotes, respectively). The first two domains, Archaea and Bacteria, include unicellular, prokaryotic organisms only. It seems that the first nucleated cells, which later gave rise to eukaryotes, were cells without mitochondria that emerged from the merging of an archaeobacterium (similar to today's *Thermoplasma acidophila*) and a eubacterium (similar to today's *Spirochaeta*). It is from such a symbiogenesis that eukaryotic cells later emerged. Some of these first cells ingested and retained oxygen-breathing (aerobic) bacteria, evolving to eukaryotic cells with mitochondria, such as those of animals. Finally, some of these aerobic cells also ingested and retained cyanobacteria and evolved to cells with plastids, such as those of algae and plants. Archaeobacteria and bacteria are generally described as prokaryotic organisms, and morphologically they do not seem to be different. However, they are very different physiologically and genetically.

Eukaryotes include both unicellular and multicellular organisms; organisms that are seemingly very diverse, such as protists, fungi, algae, plants, and animals, belong to the same domain. This should not be surprising, given that most organisms on Earth are unicellular. What is more interesting is that at the level of proteins, archaeobacteria are more similar to eukaryotes than to bacteria. Counterintuitive as it may be, eukaryotic, complex, multicellular organisms like ourselves seem to be more related to some prokaryotes than these are to other prokaryotes. This entails that eukaryotes evolved from archaeobacteria. One would then expect that eukaryotic DNA sequences similar to bacterial ones would only exist within mitochondria and chloroplasts. However, this is not the case, as DNA sequences similar to bacterial ones have also been found in the nuclei of eukaryotes.

This suggests that the evolution of eukaryotes has not been a vertical process of descendance, but involved extensive exchange of genetic material between cells, described as horizontal DNA transfer. Let me explain how this works. When a bacterium reproduces, it divides by a process called binary fission, and gives rise to two bacteria that should be genetically identical to each other and to the maternal cell (unless some kind of mutation took place). However, during their cell cycle bacteria can exchange plasmids (small, circular DNA molecules that are distinct from their main circular DNA molecule that makes up most of the genetic material of bacteria) and consequently DNA sequences e.g., such as those conferring resistance to antibiotics. Thus, if there are two strains of bacteria – e.g., the ABCD and the EFGH ones, where A–H are DNA sequences related to some cellular process – binary fission would only produce cells with these genetic structures. However, exchange of DNA molecules among bacteria (e.g., plasmids) allows the emergence of new genetic combinations (e.g., ABCDG or EFGHD, etc.). Thus, after several generations of extensive transfer of DNA molecules from cell to cell, numerous genetically different bacteria could emerge. Therefore, genome evolution in prokaryotes is not entirely tree-like; it is best represented by a complex network that resembles the branches of a tree with numerous horizontal connections.

An important point to note is that there is no way we can have the "real" tree of life – in other words, to accurately recapitulate the exact path of the evolution of life on Earth. The representation that will emerge each time will depend on the data used (DNA or protein sequences) to produce it, and this is why as more genomic data become available the details will change. Nevertheless, the fundamental similarities among organisms are crucial evidence for the common ancestry of all life. Perhaps the most convincing one is the genetic code – in simple words, the way in which RNA sequences are translated into proteins through a particular correspondence between triplets of RNA nucleotides and amino acids. The genetic code is practically universal among all extant life forms, despite some deviations in organelles and prokaryotic organisms with small genomes. That there is a common manner to "read" the "information" in the DNA of all organisms, with some deviations, is best explained by the common ancestry of all life on Earth.

Homology and Common Descent

When two characteristics are related through common ancestry, they are described as homologous. However, this kind of relation does not entail that homologous characteristics are similar. Homology may not be evident in structure but may exist at a deeper (molecular or developmental) level. Therefore, apparently homologous structures can be formed by different developmental paths, and the development of apparently non-homologous structures may be under homologous genetic control. For example, all tetrapods (vertebrates with four limbs) have digits in their limbs. Digits are considered as homologous characteristics among tetrapods; however, the developmental processes that produce them may differ. In all tetrapods, except salamanders, digits separate from each other during embryonic development as the result of apoptosis, a process of programmed cell death that creates inter-digital spaces. In salamanders, however, it is not apoptosis but the differential growth of the digits that produces them. Thus, whether or not the digits of tetrapods are considered as homologous depends on the definition of homology that is being used. In addition, different structures can be under homologous genetic control: butterflies, flies, and beetles are winged insects having two pairs of dorsal appendages that are homologous, in the development of all of which the *Ubx* (Ultrabithorax) DNA sequence is implicated. These appendages are the forewings, which are flying organs in flies and butterflies but protective organs in beetles, and the hindwings, which form functional wing blades in butterflies and beetles but are sensory organs (halteres) in flies.

Given these considerations, we can define homology as a relation of *sameness* between two or more characteristics in two or more organisms. Sameness is not simply about similarity of structure or function, but implies a historical continuity through evolution. Therefore, homologous characteristics are those that derive from the same characteristic in the most recent common ancestor of those organisms. Having been explicit about the importance of microbial life in the previous section, for the purpose of clarity and comprehensibility in this chapter I use examples of homologies in vertebrates. All people are familiar with vertebrates and so you will be able to observe on your own some of the characteristics discussed here if you go to a zoo or even while you are eating your fish or chicken. Furthermore, humans are vertebrates and it is important to realize the enormous sameness between ourselves and other animals. Hence, the question is: What inferences can we make from sameness? Recall the comparison of sharks and dolphins in Chapter 3. They both have hydrodynamic shapes that certainly facilitate swimming underwater, although they significantly differ in other characteristics, i.e., in how they breathe. What is the conclusion about the relatedness of organisms who share similar characteristics?

Vertebrates are traditionally divided into seven major groups: jawless fish, cartilaginous fish, bony fish, amphibians, reptiles, birds, and mammals. (There are different views on how these groups should be classified; I have chosen a popular classification here, so that the reader is not lost in the details, such as that birds are not considered as a distinct class from reptiles). There are various ways in which these groups can be compared to and distinguished from one another: limbs/no limbs; hair/no hair; mammary glands/no mammary glands; feathers/no feathers; lungs/no lungs, etc. If we choose one characteristic only, e.g., limbs, we only manage to distinguish between two groups each time: those having and those not having limbs. However, depending on their conception of each vertebrate class, two people might come up with different topologies – that is, different patterns of branching. For instance, if one had in mind snakes as the exemplar of reptiles, then these would be classified among vertebrates without limbs. But if one had in mind crocodiles as the exemplar for reptiles, then the latter would be classified among vertebrates with limbs. It therefore seems that we face two kinds of problems here: (1) which characteristic(s) should one choose to construct evolutionary trees; and (2) how many characteristics are adequate? The answer to the first question is that homologies, common characteristics derived from a common ancestor, are the appropriate ones to use for constructing evolutionary trees. These could be either morphological characteristics or DNA sequences. The answer to the second question is that the more characteristics we use, the better it is. The more data are available, the more accurately we are able to group the various taxa, even though the kind of data and the tools of analysis matter a lot.

One might also think that vertebrates vary significantly as, e.g., fish are very different from mammals because they lack limbs, lungs, and other major mammalian characteristics. However, most of the major anatomical characteristics of all vertebrates, including humans, existed in fish. The skeletons of the various vertebrates exhibit significant similarities and are considered to be homologous. It seems that about 90 percent of the human anatomical structure was formed during the Devonian period, some 380 million years ago. Figure 5.2 provides a visual confirmation of this. In this figure, a human skeleton (left) is compared to the skeleton of the imaginary "Gogonasus man" (right). The latter consists of the bones of a Devonian advanced lobe-finned fish that are also present in the human skeleton. These bones have been drawn to the same scale as the human ones. It seems that the lack of digits is the only major difference; the evolution of vertebrates otherwise includes rearrangements of the same basic skeleton. This example shows that organisms may actually be less different than what we usually think.

Let us now consider the concept of homology in more detail. Homologies are similar characteristics, structures, properties, processes, modules (distinguishable, partially independent, interacting units, such as segments), or sequences, which are derived from a common ancestor and which are

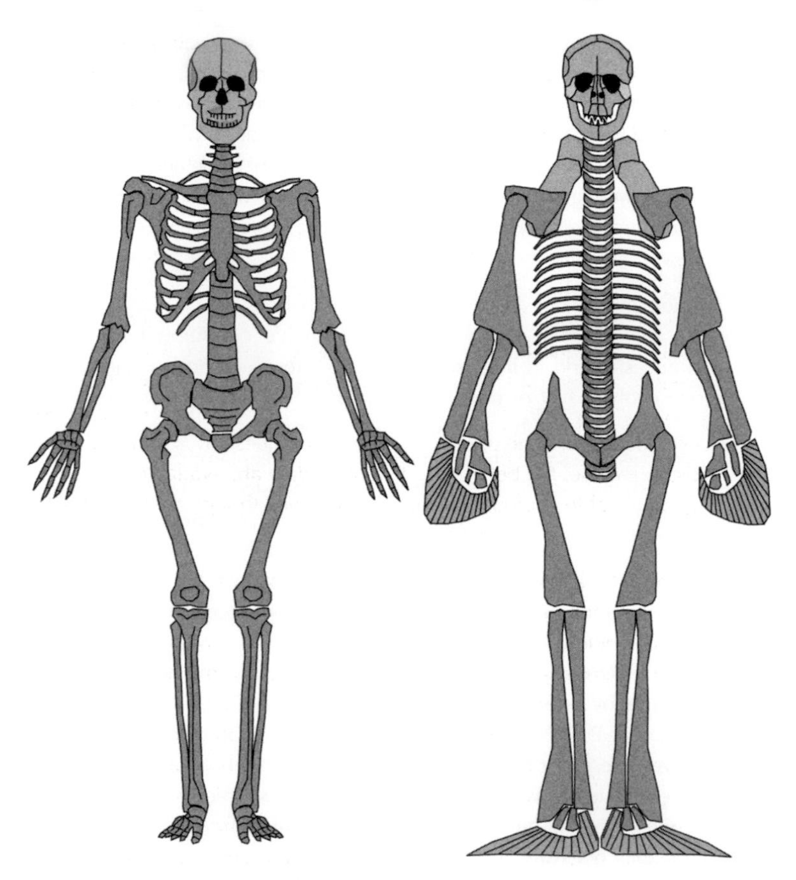

Figure 5.2 The skeleton of a modern human compared to the skeleton of the imaginary Gogonasus man, which consists of the bones present in both humans and Devonian advanced lobe-finned fish, drawn to the same scale as humans.

common among the members of a taxon. If these currently are in the same primitive condition in which they are also found in the common ancestor, they are called plesiomorphies; if they currently are in a different, derived condition, then they are called apomorphies. Thus, homologies can be shared characteristics in ancestral, or plesiomorphic, form – in this case they are called symplesiomorphies; or they can be shared characteristics in derived, or apomorphic, form – in this case they are called synapomorphies. For example, the feather, considered as an epidermis derivative, is a bird apomorphy within the clade of amniotes, whereas it is a plesiomorphy within the clade of birds. To give another example, the wings of birds and bats can be considered homologous as tetrapod forelimbs (and in this sense they are synapomorphies), but they are not homologous as tetrapod wings.

Among all these, it is synapomorphies that are most useful for phylogenetic classification and the construction of evolutionary trees. If a characteristic is in a plesiomorphic form, i.e., in the same primitive condition both in the common ancestor and in its descendants, no inference can be made about which one evolved first or how closely related the descendants are to one another. The reason for this is that both the common ancestor and its descendants possess the same characteristic. For example, lampreys, sharks, and trout are all fish that lack limbs, as the common ancestor of all vertebrates did. In this case, the lack of limbs (which is a plesiomorphy) provides no information about which of these taxa are more closely related. In contrast, characteristics in apomorphic form can be informative because a shared apomorphy (synapomorphy) suggests derivation from a common ancestor in which the apomorphic characteristic first appeared. As is illustrated in Figure 5.3, lizards, eagles, and hens all have four limbs, which is a characteristic also shared by a common ancestor that might look like a lobe-finned fish such as *Tiktaalik*. Possessing limbs is an apomorphic characteristic, given that the primitive state is not having limbs. Then, eagles and hens have forelimbs which have evolved to wings, a characteristic also shared by their common ancestor, which might be a feathered dinosaur like *Archaeopteryx*. (Note: Neither *Tiktaalik* nor *Archaeopteryx are* the common ancestors!) But how do we know whether it was forelimbs that evolved to wings or wings that evolved to forelimbs? The answer is generally given by outgroup comparison; in the case of features of the vertebrate skeleton, this can be supported by fossil evidence. Lobe-finned fish like *Tiktaalik* are estimated to have lived approximately 380 million years ago, whereas feathered dinosaurs like *Archaeopteryx* are estimated to have lived approximately 150 million years ago. So, forelimbs evolved to wings.

Unfortunately, fossils are not available for all groups of organisms. Their use in reconstructing phylogeny is very limited outside vertebrates and a few

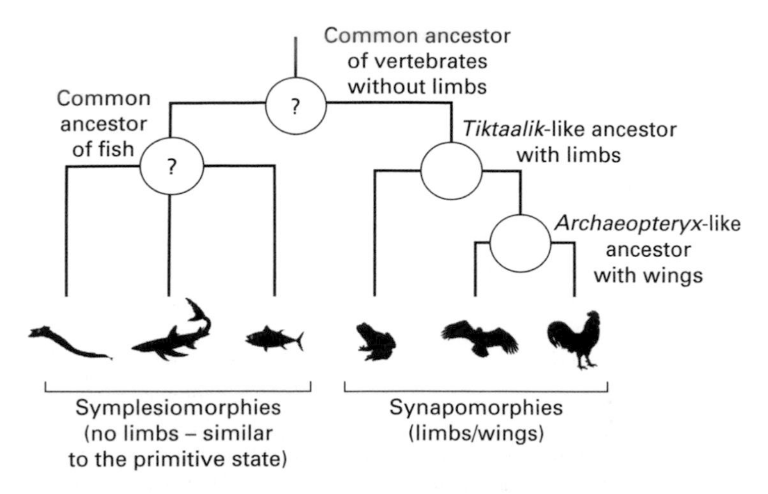

Figure 5.3 Symplesiomorphies and synapomorphies, or how the study of form of extant taxa as well as that of extinct taxa in dated fossils can help clarify evolutionary relatedness. Note: *Tiktaalik* and *Archaeopteryx* are *not* the common ancestors; however, the common ancestor in each case could be similar to *Tiktaalik* or *Archaeopteryx* and so these are included in the evolutionary tree.

invertebrate taxa. When they are available they provide information about extinct life forms, and serve as points of reference in order to establish the chronology of events. Fossils can provide important information about characteristics of extinct taxa, divergence times for taxa, times of appearances of homologous characteristics, times of extinction, or some extinct genomes. In contrast, fossils cannot provide a full array of characteristics of fossil taxa, unequivocal ancestor–descendant links, or very ancient genomes. Fossil data can thus be combined with molecular data to complement each other in calibrating evolutionary events. Whereas the fossil record is imperfect and molecular clock methods are uninformative on their own, they can be used together to construct reliably dated trees.

To make clearer how we can reconstruct the chronology of past events, here is a thought experiment. Imagine you have several white, soft balls with a sticky surface on which small pieces of paper can very easily be attached when the balls roll over them. Imagine also that you have two intersected, curve-shaped slides on which you have put small pieces of paper with different colors (corresponding to each of the slides, black and gray, respectively) and numbers (corresponding to the vertical distance from the point the balls were released, say 1–5). When you release balls consecutively, each of them will follow different routes. Suppose you release four balls and get the following pieces of paper on each:

a. black–1, black–2, black–3, black–4, black–5
b. black–1, black–2, gray–3, gray–4, gray–5
c. gray–1, gray–2, gray–3, gray–4, gray–5
d. gray–1, gray–2, black–3, black–4, gray–5

From the color and the number of the pieces that will be found on each ball, you can infer the exact route of each ball on the slides, as you can know where exactly each of the pieces of paper was initially put. In other words, from the "present" characteristics of the balls (which pieces of paper are found on each one of them) and from their "history" (where on each slide each piece of paper was initially put, and in what order compared to the other papers) we can infer the route taken. In the same sense, from the present (apomorphic) characteristics of taxa and from their first appearance in the fossil record, we can infer the evolutionary history of taxa and consequently their relationships.

In order to construct an accurate evolutionary tree of vertebrates, we should not rely on a single characteristic only, e.g., the existence/absence of limbs, but on more than one apomorphic (derived) characteristic. We could thus construct a table like Table 5.1, which includes several such characteristics. Based on this table it is possible to construct the evolutionary tree in Figure 5.4. It is important to note that the evolutionary tree can be drawn in different ways, as shown in Figure 5.5. All trees in that figure are equivalent. In human family trees it is usually the case that older offspring appear leftmost, whereas their youngest siblings appear rightmost. This is not the case for evolutionary trees, where it does not matter which group is on the left or on the right, or whether two groups are horizontally close to each other or far apart. Consequently, the apparent progression from one group to another, such as of jawless fish to mammals in Figure 5.4 (from left to right), is just an illusion.

Evolutionary scientists study not only morphological characteristics, but also DNA, RNA, or protein sequences. The rationale for constructing evolutionary

	Apomorphic characteristics						
Taxonomic groups	**Jaws**	**Bone skeleton**	**Four limbs**	**Astragalus bone**	**Diapsid skull**	**Wings**	**Synapsid skull**
Jawless fish							
Cartilaginous fish	✓						
Bony fish	✓	✓					
Amphibians	✓	✓	✓				
Reptiles	✓	✓	✓	✓	✓		
Birds	✓	✓	✓	✓	✓	✓	
Mammals	✓	✓	✓	✓			✓

Table 5.1 Some derived characteristics among vertebrates (not all of these groups correspond to classes). Such tables can be constructed by using DNA, RNA, or protein sequences and used for inferring evolutionary relationships

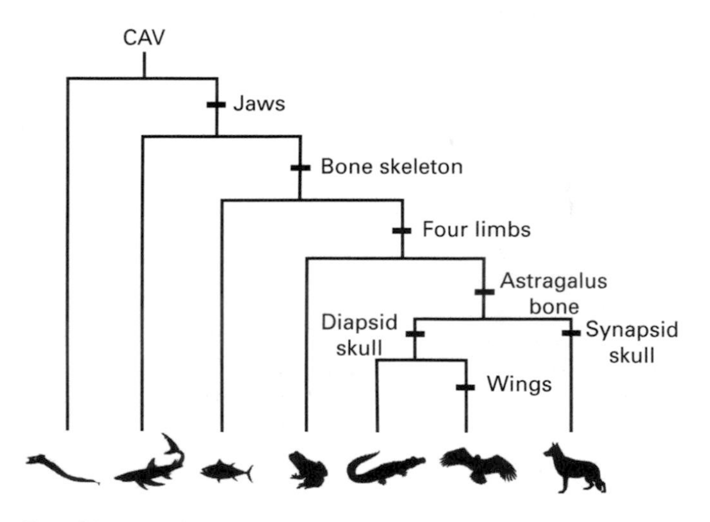

Figure 5.4 An evolutionary tree of vertebrates based on apomorphic characteristics (CAV: common ancestor of vertebrates). Note that the apparent progression from jawless fish to mammals (from left to right) is an illusion (see next figure).

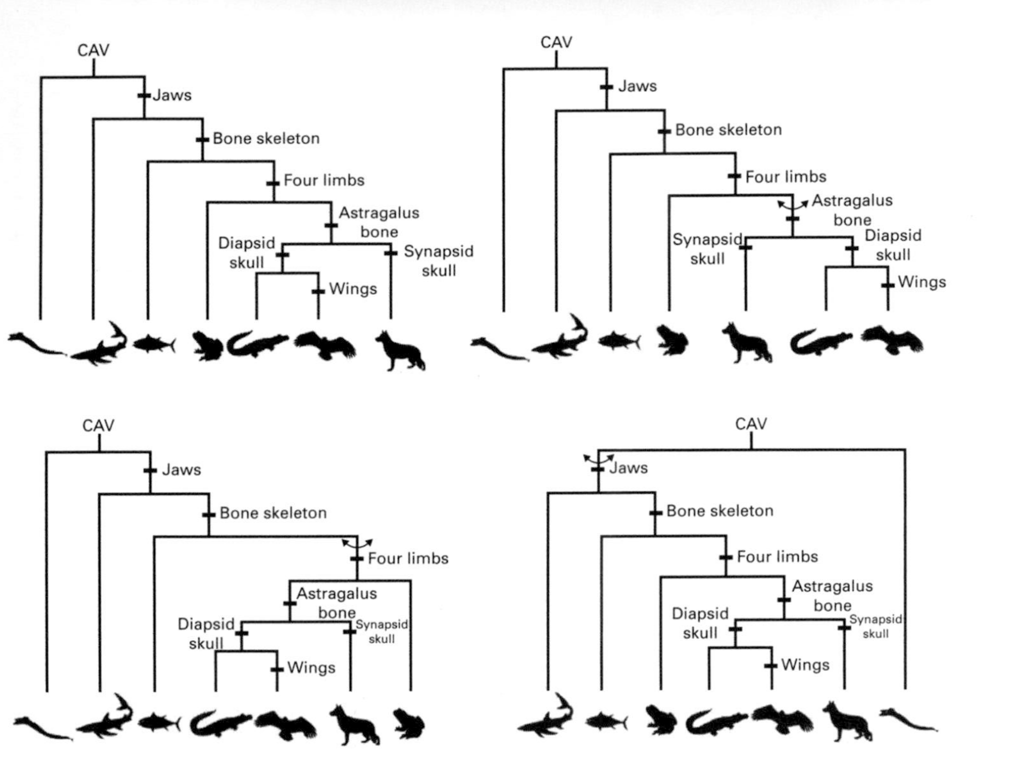

Figure 5.5 The phylogenetic relationships of vertebrates of Figure 5.4 can be represented in various ways, as shown here. All these evolutionary trees are equivalent and represent the same relationships (the double arrows indicate the nodes in which the taxa belonging in a clade have just changed relative places; CAV: common ancestor of vertebrates).

trees with DNA sequences is more or less the same as the one that relies on morphological characteristics. Fossils are used again to confirm or reject hypotheses about evolutionary relationships. Nowadays, with molecular data being accumulated at a fast pace, scientists rely on homologous DNA sequences to reconstruct several different kinds of phylogenies. However, this is not always possible, because phenomena such as horizontal DNA transfer can blur the whole picture. Another reason for this is our difficulty in distinguishing homologies from those characteristics described as homoplasies, to which we now turn.

Homoplasy and Convergence

Similar characteristics are not always due to common descent. It can be difficult to distinguish between similarity and sameness – in other words, between simply similar characteristics and similar characteristics derived from a common ancestor. Let me illustrate this. Figure 5.6 depicts an imaginary "evolutionary tree of names." "Jonathan" is the primitive state of the characteristic. This can gradually "evolve" to "Nathan" but also to "Jon." However, "Nathan" can also "evolve" from Nathaniel. Thus, different "Nathan" states can be derived from a common ancestor or may have "evolved" independently, and these are described as homologies or homoplasies, respectively. There are two important questions here. The first is how one can decide whether two "Nathan" states are homologies or homoplasies. The second is whether taxa "Jonathan" and "Nathaniel" are closely related or not, and whether the "nathan" suffix of "Jonathan" and the "nathan" prefix of "Nathaniel" could in fact be a homology at some deeper level (by the way, this presentation implies nothing about the names themselves, which are only used here for illustration purposes).

Keeping this example in mind, let us now turn to organisms and try to answer these questions. In the previous section, I mentioned that bird wings and bat wings are homologous as tetrapod forelimbs, but are not homologous as tetrapod wings. This means that both kinds of wings when considered as forelimbs, that is, only in terms of structure and not in terms of function, are homologous since they were derived from their common ancestor. However, when their function is also considered, one can no longer talk about homologous structures since the use of bird and bat forelimbs as wings for flight is not due to their being derived from their common ancestor. In contrast, the

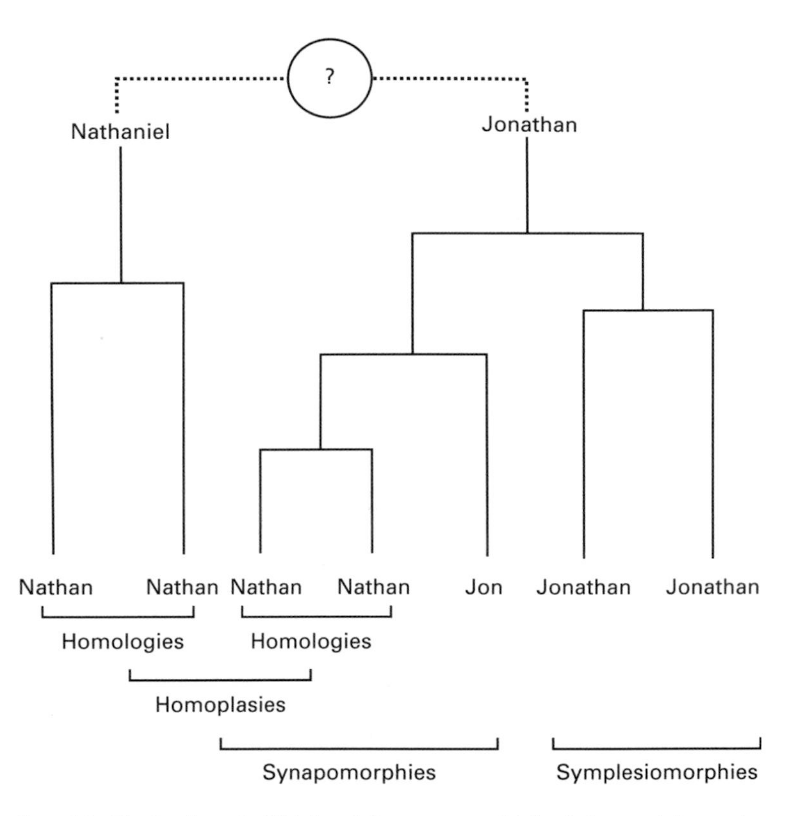

Figure 5.6 The "evolution" of "Nathan." Contemporary "Nathan" characteristics can be homologies if they have evolved from a common ancestor (either "Nathaniel" or "Jonathan"), or homoplasies if they have evolved independently from different ancestors ("Nathaniel" and "Jonathan"). "Nathan" and "Jon" are synapomorphies, i.e., derived characteristics from an ancestral one ("Jonathan"), whereas contemporary "Jonathan" are symplesiomorphies.

evolution of bird and bat forelimbs to wings took place independently from each other through a phenomenon described as convergence. Characteristics like these, which evolved independently and which are not derived from a common ancestor, are called homoplasies (Figure 5.7).

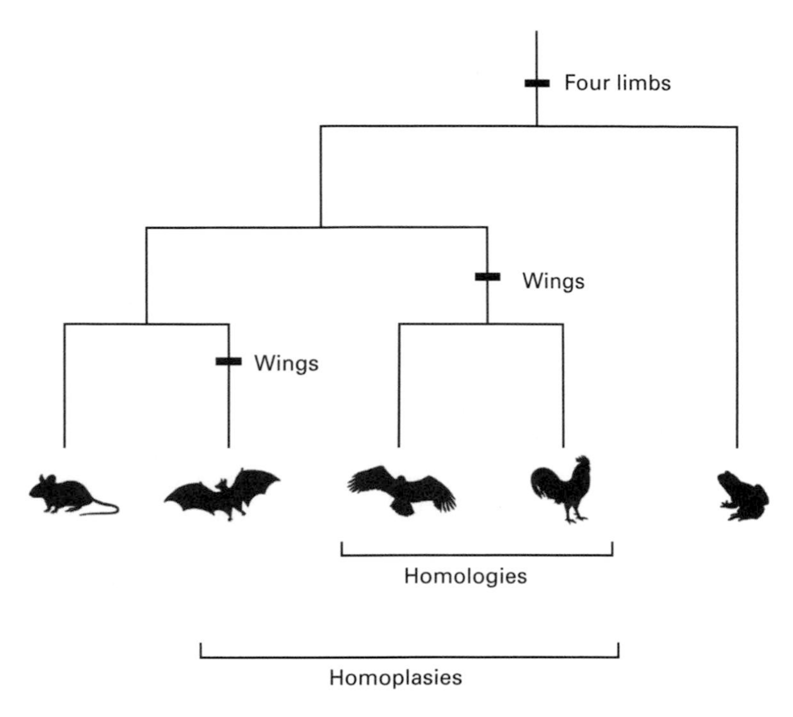

Figure 5.7 Eagle and hen wings are homologies as they are derived from a common ancestor; both of these and bat wings are homoplasies.

Homologies and homoplasies have been considered as antithetical concepts. One may intuitively think of homologies as the product of evolutionarily conserved developmental processes and homoplasies as the product of independent ones. However, this is not actually the case. It seems that there is a continuum of biological processes from one to the other, including parallelism and convergence. Similar developmental processes may produce different structures, but also different developmental processes may produce similar structures, even in distantly related organisms (see, respectively, the examples of insect dorsal appendages and tetrapod digits in the previous section). Since homology has been defined here as a relation of sameness between two or more characteristics that derive from the same characteristic in their most recent common ancestor, homoplasy can be defined as a relation of similarity between two characteristics in two or more organisms that are not derived from the same characteristic in their most recent common ancestor.

Let us consider convergence because it is the clearest case of homoplasy and as such can be clearly distinguished from homology. Convergence refers to the emergence of the same characteristic through independent evolution, i.e., from different ancestral characteristics. The main difference between convergence and parallelism is that whereas convergence refers to the evolution of the same characteristic, C, from two different ancestral characteristics, A and B, in two different lineages, L1 and L2, parallelism refers to the emergence of C from the same ancestral characteristic, A, in two different lineages, L1 and L2. In the case of bat and bird wings the forelimbs evolved to different kinds of wings. The wings of bats consist of their elongated digits, which are connected via a webbed membrane of skin, whereas the wings of birds consist of their whole forelimb that is covered by feathers. The wings of bats and birds are therefore homoplasies.

Another interesting case of homoplasy is the hydrodynamic body shape of sea mammals, such as dolphins and whales. As already discussed in Chapter 3, sharks and dolphins have the same hydrodynamic shape. However, this is not due to common ancestry, because dolphins evolved from a tetrapod ancestor. Whereas the lack of limbs and the hydrodynamic shape was a characteristic common among fish groups, most of the other vertebrate taxa have limbs. However, some mammals turned from life on land to life in the sea and evolved hydrodynamic shapes. The fossil record provides adequate evidence about how this transition could have been possible. Thus, the lack of limbs can be considered as a homology between sharks and trout, or between whales and dolphins. However, the hydrodynamic shape of sharks and dolphins is a case of homoplasy, not homology. Although both of them are vertebrates and, as such, share common ancestors, their shape is not derived from them. In the case of whales and dolphins it has evolved independently from an ancestor with four limbs (Figure 5.8). The evolutionary tree of Figure 5.7 can now be enriched with more examples of homologies and homoplasies (Figure 5.9).

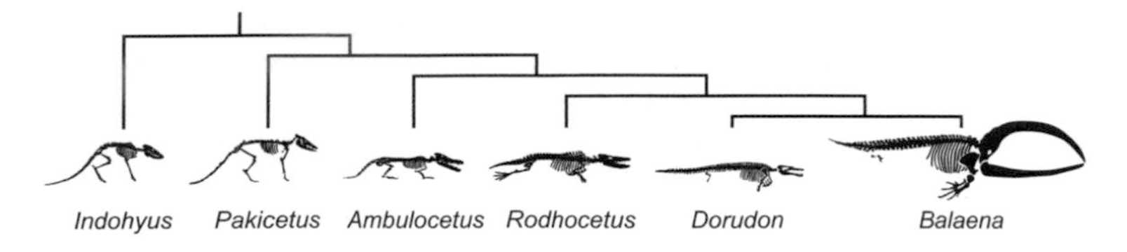

Figure 5.8 The evolution of modern whales (*Balaena*) from tetrapod ancestors.

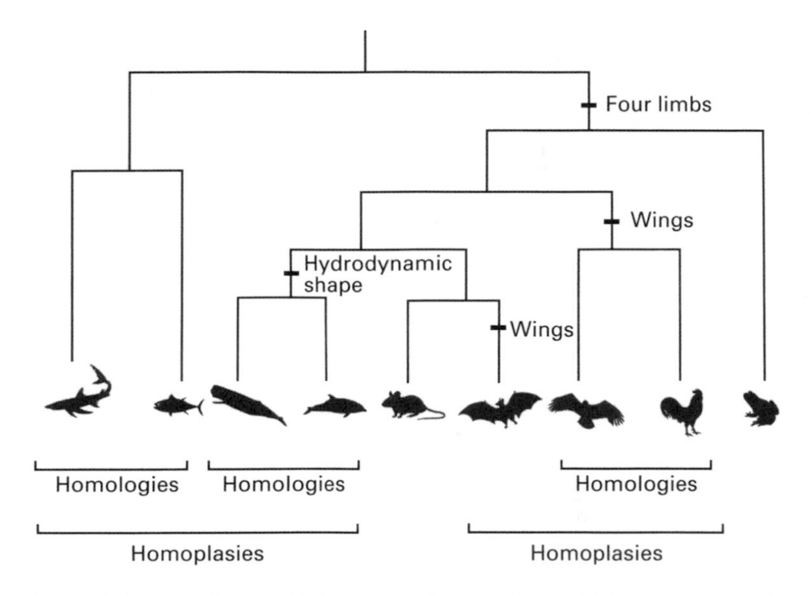

Figure 5.9 Wings of bats and birds and hydrodynamic shapes of fish and sea mammals as homoplasies and homologies.

In the case of vertebrates, scientists can rely on the fossil record to infer phylogenies and reconstruct the evolution of each taxon. However, there are cases where this can be very difficult, especially when only data from the present (e.g., DNA sequences) and not from the past (e.g., fossils) are available. Homoplasy, in general, and convergence, in particular, can be confounding factors of genuine homology in evolutionary studies. Molecular characteristics (e.g., DNA or protein sequences) typically have a few alternative states. Consequently, the probability that different species acquire the same nucleotide or amino acid independently is significant and can confound evolutionary history. Thus, one cannot simply claim that by comparing the DNA sequences of organisms we can infer evolutionary history. Are all similarities observed due to common descent or due to convergence – in other words, are they homologies or homoplasies? One strategy to overcome this problem has been to compare rare genomic changes. These are changes due to rare mutational events, such as insertions and deletions in coding sequences, which are less likely to occur independently in the same way. However, phenomena such as horizontal DNA transfer can still complicate the picture. Figure 5.10 shows alternative evolutionary trees based on DNA coding sequences and rare genomic changes, showing that in some cases we cannot know the actual evolutionary history, mostly due to homoplastic events. As shown in Figure 5.10, rare genomic changes are more informative than coding DNA sequences. A challenge that is more difficult to cope with is, as also shown in Figure 5.10, that the older the lineages are, the more homoplastic events may have taken place, and this might make the actual history more difficult to discern. The exact relatedness between chordates, arthropods, and nematodes seems to be less certain than that between gorillas, humans, and chimpanzees because evolutionary events in the former case are much older than those in the latter, and so more homoplastic events may have taken place.

This brings us to another question, which is perhaps the most difficult one to answer. Can two homoplasies exist due to a homology at some deeper level? In the first section of this chapter I described how all living forms are related through common ancestry. If all organisms are more or less related, can we actually talk about an independent evolution of characteristics? Even if we know, as in Figure 5.6, that two of the "Nathan" states have evolved from an ancestral "Nathaniel" state and the other two have evolved from a "Jonathan," are we sure that "Nathaniel" and "Jonathan" are not related through a shared common ancestor in the deep past? An example of this kind is the evolution of eyes, which were long thought to be the outcome of evolutionary convergence. More recent studies have revealed the existence of shared genetic networks for eyes in otherwise different animal taxa. It has been found that a particular set of transcription factors (proteins involved in the

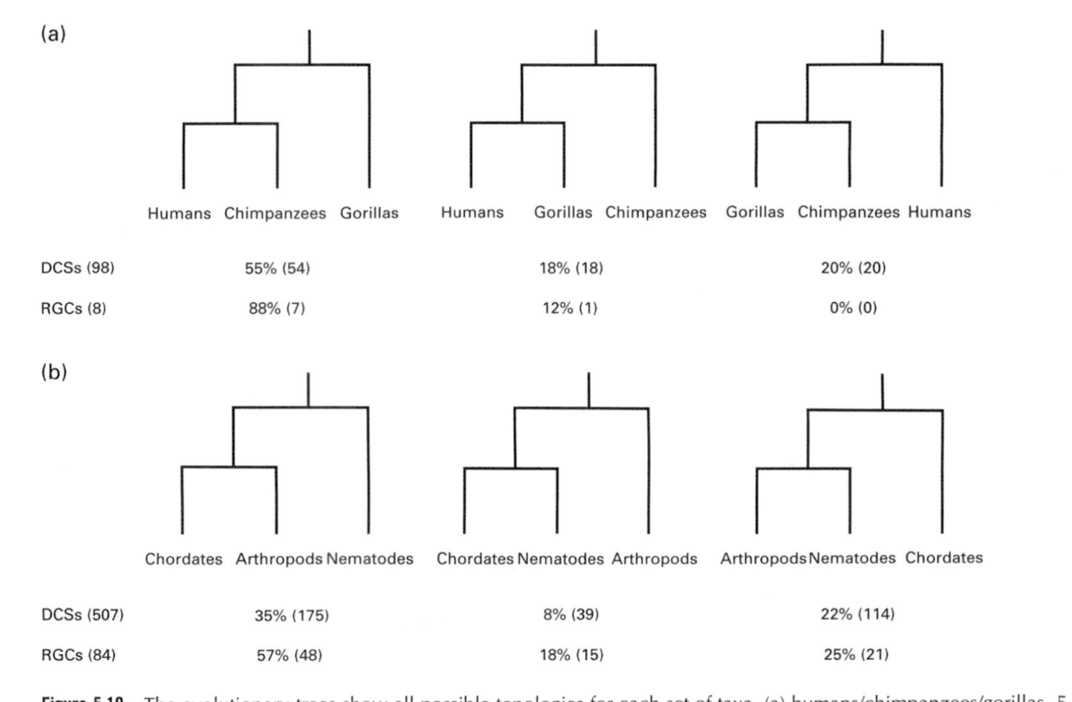

Figure 5.10 The evolutionary trees show all possible topologies for each set of taxa. (a) humans/chimpanzees/gorillas, 5–8 million years ago; (b) chordates/arthropods/nematodes, >550 million years ago. Below each topology are the percentage and number (in parentheses) of DNA coding sequences (DCSs) and rare genomic changes (RGCs) supporting that topology (based on maximum likelihood analyses). A number of DCSs in each case are uninformative: (a) 6 of 98 DCSs; and (b) 179 of 507 DCSs. What the resulting trees indicate is that we cannot be certain about which topology is the accurate one. However, RGCs increase the level of certainty compared to DCSs.

expression of DNA sequences), such as those produced by members of the *eyeless, atonal,* and *eyes absent* families of DNA coding sequences in *Drosophila melanogaster* and their homologues in vertebrates, are involved in the specification and formation of various types of animal eyes. In addition, the ability to detect light in all light-sensing organs in animals depends on a set of chemical reactions that involves opsin proteins, and so it has been assumed that all modern variations of light sensing in bilaterians (organisms with twofold symmetry that gives them definite front and rear, as well as left and right, body surfaces) can be traced to the existence of photosensitive cells in a common ancestor with PAX6 and other transcription factors involved in a regulatory pathway leading to opsin production. This phenomenon, in which the development and evolution of morphologically disparate organs depends on homologous genetic regulatory circuits, has been described as *deep homology*.

Let us consider multicellularity, the property of an organism to consist of many cells. How can this be explained? Is this due to some deep homology, because unicellular organisms have the inherent tendency to form multicellular ones? Or is it a case of homoplasy and multicellular states are selectively favored and evolve once they arise? To answer these questions, we first need to distinguish between two kinds of multicellularity: simple multicellularity and complex multicellularity. Simple multicellular organisms have the form of filaments, clusters, balls, or sheets of cells. Although they may have differentiated somatic and reproductive cells, they do not exhibit more complex patterns of differentiation. Simple multicellular eukaryotes consist of cells connected to each other by adhesive molecules, but there is not much communication or transfer of resources between cells. In contrast, complex multicellular organisms exhibit both cell-to-cell adhesion and intercellular communication, as well as tissue differentiation mediated by networks of regulatory DNA sequences. Simple multicellularity has evolved many times among the eukaryotes, but complex multicellular organisms belong to only six groups: animals, land plants, two groups of algae, and two groups of fungi. It seems that at least in animals and plants multicellularity evolved with adhesion, which gave rise to simple multicellular structures from a single progenitor cell. The next step seems to have been the evolution of bridges between cells, which facilitated the transport of nutrients and signaling molecules between cells, something that is found in all groups with complex multicellularity, but not in others. It seems that further evolution of multicellularity involved both co-optation and *de novo* evolution of signaling molecules – molecules involved in transmitting information between cells – and transcription factors, which led to more complex body structures. The transition to multicellularity seems to be facilitated by "ratcheting" mutations that stabilized the multicellular state and allowed the evolution of multicellular complexity.

Homoplasy confounds evolutionary histories and what has long been explained as the outcome of convergence may actually be the outcome of homology at a "deeper" level. This does not mean that we cannot know the evolutionary histories of particular taxa, or that our evolutionary explanations are not reliable as we cannot definitely distinguish homology from homoplasy. What it means is that there will always be uncertainties. One way to deal with this could be to distinguish between different levels. It could be the case that convergence occurs at the level of the organism due to the fact that particular individuals survive and reproduce in a particular environment. But, at the same time, some kind of homology can exist at a deeper level, as the organismal forms that are being selected differ in their developmental processes from those that are not. In the case of the evolution of vertebrates, which has been the focus of my discussion so far, light can be shed by the very active research field of evolutionary developmental biology, which is the focus of the next section.

Evolutionary Developmental Biology

So far, I have focused on vertebrates – which include humans – because it is their evolution that many people find difficult to understand. In Chapter 3 I noted that artifacts have more fixed essences than organisms. Here, I explain why the essences of organisms are not fixed due to the interrelation of evolution and development. Such phenomena are studied by evolutionary developmental biology (usually dubbed as evo-devo). Evo-devo studies both the evolution of development (i.e., how developmental processes evolve) and the developmental basis of evolution (i.e., how development shapes the evolution of organismal characteristics). In this section I explain some significant features of the evolution of multicellular organisms in order to show both how complexity is possible as the outcome of natural processes without any assumption of design, and how large morphological changes in evolution are possible due to – sometimes minor – changes in the underlying developmental processes.

The focus of most public discourse about evolution has been on whether natural selection, Darwin's main principle, is adequate to drive evolution: the divergence of populations of the same species that are modified and eventually become new species descending from a common ancestor. In a later book, Darwin used a metaphor to illustrate the process of selection. He contrasted the conscious selection of stones made by a builder who is using them to build a house with the accidental variation of their shapes and sizes. Variation is a prerequisite for selection, as the builder would not be able to make any selection if all stones were the same. But, beyond this, according to Darwin, variation does not affect the outcome. It is selection that guides the outcome as the builder selects the stones appropriate for the building. In the same sense, Darwin suggested that it is selection that drives evolution in nature. Since then, and in many cases until today, evolution has been described as a two-step process, involving the "random" emergence of variation and the "non-random" process of selection.

This is where a crucial obstacle in understanding evolution arises. Let us assume that selection is a natural process that can produce adaptations by accumulating favorable variations. How is it possible to randomly produce the appropriate variation for this process? To use Darwin's own example, let us assume that the builder is competent enough to build a house from stones. What guarantees that the stones fallen from a precipice will be appropriate for this? What if the small stones fall first and the big ones follow much later? Can one build a stable house by putting large stones on small ones? And what if only small stones are available? Will the builder be able to build a stable house with small stones only? Darwin's metaphor is not a good one for two reasons. First, the conscious and skillful builder is not a good analog of natural selection, which is an unconscious and unintentional process. Second, and most importantly, variation is neither random nor as limited as the stone example might imply. Organisms certainly have an inherent potential in their genetic material that drives the development of some characteristics but not of others. This at first sight seems to limit the potential of evolution. However, this is circumvented by the fact that variation is not as limited as the stones example indicates. In Darwin's examples, stones are stones and they differ only in size. In contrast, in organisms particular DNA sequences involved in forming a particular structure or affecting a particular process can eventually be "co-opted" to form a different structure or to affect a different process.

Let me use another example to illustrate this, before turning to a description of these processes in scientific terms. Imagine that someone opens a shop selling pizzas. The first pizzas produced are only made with dough (flour and water), cheese, and tomato. Gradually, more kinds of pizzas are produced, with sausage, onions, bacon, garlic, and several different types of cheese. But the shop is still selling pizza. However, the owner decides that she also wants to sell cookies. She thus starts using the same dough used for pizza to make cookies, with the addition of sugar and eggs. Pizzas and cookies do not look similar and taste very different, but they have the same basis: They are made of dough and several other materials. You might think of the processes of making cookies and pizzas as distinct developmental processes, where pizzas and cookies "develop" by molding the various materials together. In this case, the "adult" forms – pizzas and cookies – are very different, but some of the materials used to produce them – such as dough – are the same. The shop now sells both pizzas and cookies, but these are not doing equally well in terms of sales. Customers can also buy pizzas elsewhere in the area, and come to the shop mostly to buy the cookies that they like very much. As customers now mostly select the cookies over the pizzas, this makes the owner of the shop start making more and more cookies and fewer and fewer pizzas, and thus gradually changes the pizza shop to a cookie shop.

What has been described in this example could be perceived as equivalent to an evolutionary process that took place due to changes in development and evolution. This is what evo-devo is about. Changes in the initial developmental process (dough + cheese + tomato → pizza) gradually produced a new developmental process (dough + sugar + eggs → cookies). Then selection favored the latter developmental process over the former and eventually the first population (we can think of the pizza shop as a population of pizzas) evolved to another one (we can also think of the cookie shop as a population of cookies). This is a simple example that I hope makes clear the core of evo-devo. The question is not whether a pizza can change into a cookie (this is how evolution is often described), but how a population of pizzas (represented by the pizza shop) can evolve to a population of cookies (represented by the cookie shop) due to a change in developmental processes (materials available are used to make cookies and not pizzas). I am aware of the limitations of this example, such as that the "evolution" of cookies is driven by the intentions of the shop owner, whereas there is no such intentionality in nature. However, I think that it shows in a comprehensible manner how changes in development can affect evolution. The crucial point is that evolution does not take place with changes in individuals. It is populations that change, but then describing evolution just as the change in the genetic structure of populations is inadequate. In contrast, relating evolution as change in the developmental trajectories of individuals, which is the evo-devo perspective, is a richer description.

The take-home message intended by the above example is that evolution in multicellular organisms proceeds not by changes in adult forms (e.g., ancestral adult A is transformed to descendant adult D), but by changes in the developmental processes that produce these adult forms (e.g., the developmental process of ancestral adult A changes to the developmental process of descendant adult D). This is possible because DNA sequences that are involved in regulating development are conserved across very different groups.

One well-known example is the *Hox* DNA sequences, which influence the development of body parts in insects and vertebrates. These are implicated in the production of transcription factors, a group of proteins that influence the expression of specific DNA sequences, determining which ones are "turned on" or "turned off." They are grouped into two clusters, the *Antennapedia* complex that comprises five DNA sequences that affect the front half of the body, and the *Bithorax* complex that comprises three DNA sequences that affect the back half of the body. The relative order of these DNA sequences corresponds to the relative order of the body parts they affect. Despite their differences, they all contain the same sequence that was called the *homeobox* (which is why the DNA sequences were later called *Hox*). Evolution occurs not only due to changes in DNA sequences that directly affect characteristics (e.g., limbs), but also due to changes in DNA sequences that are implicated in the development of these characteristics. The question, then, is not only which DNA sequences an organism has, but also how their expression is regulated: when, where, and for how long it is activated and with what kind of outcome.

Let us consider two cases already discussed in the previous section: the loss of limbs in whales and dolphins and the evolution of wings in bats. Modern cetaceans, such as whales and dolphins, are characterized by the absence of hind limbs. Hind limb development is initiated in the embryo, but is not maintained due to the absence of *Hand2*, a regulator sequence that affects initial limb outgrowth in amniotes. Thus, it seems that the initial reduction in hind limb size was driven by changes in regulatory control sequences that affect development. In the case of bat wings, the digits in bats are initially similar in size to those of mice during embryonic development, but then bat digits lengthen enormously. It seems that bone morphogenetic protein 2 (Bmp2) can stimulate cartilage proliferation and differentiation and increase digit length in the bat embryonic forelimbs as its expression is increased in bat forelimb embryonic digits compared to mouse or bat hind limb digits. This affects developmental elongation of bat forelimb digits, and probably their evolution. In both of these cases, it is the change in the expression of particular DNA sequences, and not within those DNA sequences themselves, that has resulted in evolution. Changes in development produce novel phenotypes that may subsequently be favored by natural selection.

There are at least four cases of changes in development that may contribute to evolution. *Heterochrony* is about the differences in the timing of developmental events. One example is the change of relative timing of egg hatching and segment formation, which are two important events in arthropod development in centipedes. In members of the Order Geophilomorpha, segment formation ends before egg hatching, whereas in members of the Order Lithobiomorpha it ends long after that. As a result, in the former group the hatchling has the full number of segments, whereas in the latter group this number increases over the course of a year after. *Heterotopy* is about differences in the location of developmental events. An example is flatfish, the head of which is partially rotated in relation to the rest of the body, resulting in both eyes being on the same side of the skull. Their embryos are bilaterally symmetrical but during development one of the eyes moves across the head and ends up on the same side as the other. *Heterometry* is about differences in the amount of activity in developmental events. An example of this is the increasing brain size in the human lineage compared to the chimpanzee lineage. Humans are roughly 1.25 times larger than chimps, but our brains are 2.7 times larger than those of our close relatives. Finally, *heterotypy* is about differences in the type of developmental events, such as the production of hemoglobin S (HbS) in humans instead of the normal hemoglobin A (HbA). All these phenomena have been described as *developmental repatterning*.

Changes in the regulation of development can produce significant changes in body structure. One of the most stunning examples is the inversion of the dorsoventral axis of arthropods and other protostomes compared to vertebrates. Most animal phyla are bilaterians, and they are divided into protostomes and deuterostomes. Protostomes are those in which the mouth develops close to the blastopore (an opening in which the cavity of the gastrula, an early stage in animal development, communicates with the exterior), whereas in deuterostomes the anus develops close to the blastopore and the mouth develops from a second opening. In protostomes the central nervous system is closer to the ventral region (front side of the organism) and the digestive system is closer to the dorsal region (back side of the organism), whereas in deuterostomes it is the other way around. This is due to an inversion in the expression of genes that determine the dorsoventral axis (which is defined by a line that runs orthogonal to both the anterior/posterior and left/right axes). The same developmental mechanism guides body formation in both protostomes and deuterostomes, and this makes the ventral region of protostomes homologous to that of deuterostomes. This is an amazing finding that shows how very different organisms can emerge from changes in the regulation of development. Protostomes and deuterostomes thus share important underlying similarities in their development, despite their enormous morphological differences.

The findings of evolutionary developmental biology not only provide further evidence for the common ancestry of all life on Earth, but also explain how divergence at the organismal level can be due to changes in regulatory DNA sequences. Thus, organisms that have many similar DNA sequences may differ significantly due to the different ways that particular regulatory sequences are switched on and off – which in turn results in the different expression (in time, place, amount or type) of their otherwise similar DNA sequences. Most importantly, there is no paradox in how organisms may have evolved from ancestors from which they seem to be significantly different. Overall, evo-devo research shows that minor changes in DNA sequences can produce large changes at the phenotypic or the developmental level, which provide the raw material for evolution – the topic to which we now turn.

6 Evolutionary Processes

Adaptation and Natural Selection

In the everyday use of the word, to *adapt* means to make something suitable for a new use or to adjust it to new conditions. Accordingly, adaptation may refer to the process of adapting something or of being adapted. A characteristic that is the outcome of such a process might also be called an adaptation. Thus, based on these definitions and on everyday experience, one could infer that biological adaptation is the process by which populations become better suited to their environment, which might consequently mean that some of their characteristics change and become suitable for new roles. These new characteristics might be called adaptations as well. Therefore, adaptation can be defined both as a process and as a feature, and indeed these are the uses of the term in evolutionary biology. In both cases, adaptation refers to the positive contribution that a characteristic makes to the survival and reproduction of its possessors (usually described as fitness). Whereas different definitions of adaptation as a process exist, they do not significantly differ from one another. Overall, adaptation has been defined as an evolutionary process for which natural selection seems to be an important factor; the differences among the various definitions have to do with how important natural selection is.

In contrast, there has been much debate concerning the appropriate definition of adaptation as a characteristic. One definition is that adaptation is a characteristic that is effective in performing a particular role and that is the outcome of a selection process because of its effectiveness in this role. Selection, and the historical process underlying it, are important for defining adaptation, and the criterion for considering a characteristic as an adaptation should be its causal history. These are the historical definitions of adaptation. A crucial distinction for this definition is between a characteristic being favorable to its possessors *and* being selected; a characteristic is not an adaptation just because it confers some advantage to its possessors, but because their ancestors were selected due to this advantage. Being beneficial could be the result of chance; therefore, this is not an adequate criterion for a characteristic being an adaptation. A characteristic can only be considered an adaptation if it became prevalent in a population due to natural selection, even if it does not currently confer any advantage. In short, according to the historical definitions, a characteristic is an adaptation only if it is the outcome of natural selection, independently of whether it currently confers any advantage to its possessors.

There also exist ahistorical definitions of adaptation. According to those, an adaptation can be defined as a characteristic that contributes to the survival and reproduction of its possessors in a given environment. In this case, the emphasis is on the current contribution of the characteristic and not on its history. Whether a characteristic had the adaptive quality from the very beginning or not is irrelevant for being considered as an adaptation; this should happen only if it is currently favored by selection. Consequently, the history of the characteristic in general, and the selection process through which it was spread in a population or species in particular, is not that important. What is important is the current advantageous contribution of the characteristic. This, of course, means that selective history is not necessary, not that it is irrelevant. In short, according to the ahistorical definitions of adaptation, a characteristic is an adaptation if it currently confers an advantage to its possessors, which gives them better chances of survival and reproduction among a specific set of alternative characteristics, and if it is consequently favored by selection over them, independently of whether there has been selection for this characteristic in the past.

Let me use an example to illustrate the differences between these two types of definitions. One may wonder under which conditions the white color of Arctic bears can be considered an adaptation. According to the historical definitions, it is an adaptation only if it has come to be the prevalent color in that population through natural selection. This means that there must have been selection *for* this color in the past – for example, because it facilitated its bearers' concealment in a snowy environment and consequently contributed to their survival and reproduction as they could attack their prey without being easily noticed. As a result, a process of differential survival and reproduction took place: The individuals with white color had an advantage compared to the others with different colors, and eventually the white color became the prevalent one in that population. In contrast, according to the ahistorical definitions, the white color of the Arctic bears can be considered an adaptation only if it currently provides its bearers with better chances of survival and reproduction compared to other individuals with different colors in the same environment, and it is thus being selected. Whatever happened in the past is not relevant; what matters is what is happening now. For instance, the offspring of brown bears that are accidentally born white and that due to this have an advantage in living in the Arctic may also have an advantage in surviving and reproducing. If these white offspring are favored by selection, even if their color does not eventually become prevalent in the bear population, their white color can be considered to be an adaptation according to the ahistorical definitions, although it emerged by chance.

Perhaps a simpler, and better, distinction is between adaptations, characteristics that exist as a consequence of natural selection for one or more of their effects, and adaptive characteristics that contribute to the survival and reproduction of their bearers, whether they are the outcome of natural selection or not. According to this distinction, adaptations do not need to be currently adaptive, whereas adaptive characteristics may also be adaptations, but they may arise by chance as well. What is more important to note, though, is that in all cases it is populations – not individuals – which adapt (in the process sense). Individual bears that were accidentally born white cannot be said to have adapted to their environments. Rather, it is the bear *population* that has adapted (past selection) or is adapted (current selection) because more white bears than bears with other colors survive(d) and reproduce(d) in the particular environment. Whether adaptation is the outcome of a historical or a contemporary process of selection, it is always a property of populations.

Now, what if a currently beneficial characteristic initially evolved for some other role or no role at all, and later became "co-opted" for its current role? It has been suggested that such a characteristic should be distinguished from adaptations and be called an exaptation. Consider two different evolutionary processes, P1 and P2 (Figure 6.1). In P1, brown color appears in a given environment and is selected because it increases the survival and

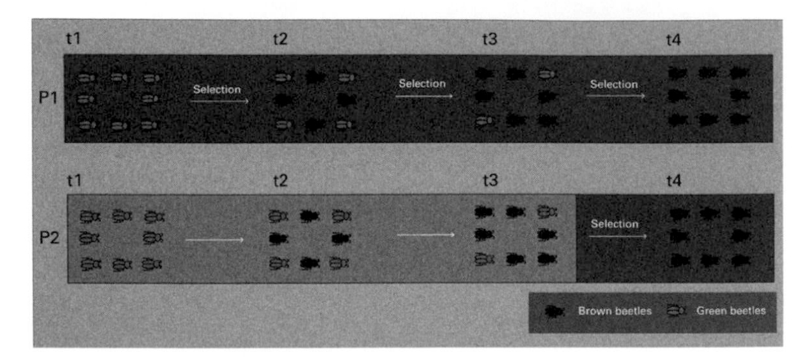

Figure 6.1 Adaptation (P1) and exaptation (P2). In P1, brown color in beetles is an adaptation because it was selected during t1–t4. In P2, brown color is an exaptation because it became prevalent in the population of beetles up to t3 without conferring an advantage, and was "co-opted" during t3–t4 (in all cases, proportions, not actual numbers of the various types of individuals, are depicted).

reproduction of its possessors. Thus, brown color is an adaptation because it was selected for conferring an advantage (concealment) to its bearers in the dark environment. In P2, brown color emerges and spreads in a population for reasons other than selection. The environment had a light color and so there was no advantage for beetles with brown color. Then, an environmental change happens – the environment gets darker – and the characteristic comes to play an important role because the brown beetles gained a concealment advantage. Such a characteristic that initially spread for non-adaptive reasons and was later co-opted for a new role in the new environment is an exaptation. A characteristic could also spread in the population due to selection for one kind of advantage and later be selected for a different kind of advantage. Exaptation does not require (though it can have) a non-adaptive origination step.

So far, I have been writing about selection *for* a characteristic. However, this is not the only kind of selection possible. Imagine a toy with balls of different sizes and colors (e.g., white balls are large and black ones are small) in which only the small ones can pass through holes when the toy is shaken, whereas the large ones cannot. In this way, there is *selection for size* (small balls pass; large balls do not pass). However, it is also the case that only black balls and

Figure 6.2 Selection *for* size and selection *of* color. White balls are large and do not pass through the holes. Only black balls pass to the bottom layer because they are the smallest. In this case, there is selection *for* size (small balls pass to the bottom layer; large balls do not) and selection *of* color (black balls pass to the bottom layer because they are also small; white balls do not because they are also large).

no white balls pass to the bottom, because all large balls are white whereas all small balls are black. In this case, there is no *selection for color*: Whether a ball is black or white makes no difference to whether it can pass through the holes. However, while there was selection *for* small size because the small balls could pass through the holes, there was also selection *of* black color that was incidental because black balls happened to have small size (Figure 6.2). In this sense, I distinguish between two kinds of selection relevant to characteristics: selection *for* a characteristic if it is that characteristic that is being selected, and selection *of* a characteristic if its selection is incidental.

Two examples that make this distinction clear are pleiotropy and linkage. Pleiotropy is the phenomenon in which a DNA sequence affects two different characteristics. Imagine a DNA sequence G that is implicated in the development of two characteristics, A and B. If there is selection *for* characteristic A, the consequence will be that more organisms with this characteristic will survive and reproduce. However, these individuals will probably have characteristic B, too. Thus, selection *of* characteristic B will be a consequence of selection *for* characteristic A. In the case of linkage, imagine that two linked DNA sequences G_A and G_B (that is, found on the same chromosome) affect the development of characteristics A and B, respectively. If there is selection *for* characteristic A, the consequence will be that organisms with this characteristic will survive and reproduce. However, these individuals with sequence G_A will also likely have the sequence G_B – because the two sequences are linked – and therefore they will also have characteristic B. Thus, selection *of* B will be a consequence of selection *for* A. In both of these cases, there is selection *of* characteristic B but nevertheless this characteristic is not considered as an adaptation because there is no selection *for* it, as was the case for characteristic A.

Speaking of selection, it is important to note the impact of the metaphors used. Adaptations can evolve through natural selection only when the latter operates for many generations on variants of characteristics, which first arise independently of the advantage they may eventually confer to their bearers. In this sense, several different adaptive variants may coexist in the population. This was Darwin's view in the *Origin*; he wrote about "the accumulation of slight but useful variations." But this is different from the idea of the "survival of the fittest," a term coined by Herbert Spencer and eventually adopted by Darwin in the fifth and subsequent editions of the *Origin*. In this case, there is no accumulation of slight, advantageous variations, but simply elimination of the individuals that bear the non-advantageous characteristics and survival of those individuals that possess the advantageous ones. Therefore, there is an important difference between the two views. Darwin's initial view of selection was one of *selection for*, whereby adaptation is the cumulative effect of selection operating on the available variation for many generations. The subsequent view, based on the idea of the survival of the fittest, is of *selection against*, whereby variation is quickly eliminated. In other words, the first view is about the gradual selection over trans-generational time *for* adaptive characteristics, whereas the second view is about selection *against* organisms that do not have such characteristics.

Understanding the difference between these two kinds of selection is a major conceptual issue, because it is difficult to understand how novel adaptations can evolve through selection if the latter just eliminates variants. Selection *for* a characteristic does not simply mean that this characteristic is preserved and that all the others are eliminated. It rather means that slight variations of this characteristic provide an advantage, and so those individuals bearing them survive and reproduce more efficiently than others. Gradually, over many generations, these variations of the characteristic become prevalent in the population. Thus, selection is not just an eliminative process; it is rather a "creative" process that accumulates useful variations in the population and gradually drives it to adaptation. Figure 6.3 provides an illustration of this difference.

Stochastic Events and Processes in Evolution

Several processes can drive the evolution of a population without any kind of selection. These are generally described as stochastic processes: undirected processes with unpredictable outcomes in which chance seems to play a major role. The main difference between these phenomena and natural selection is that whereas the direction of natural selection can be predictable if the antecedent conditions are known, the direction of stochastic processes and the eventual outcomes are largely unpredictable. There are different kinds of stochastic processes at various levels, but perhaps the most important ones are those at the molecular level. The reason for this is that even if we accepted that natural selection is the only process that can lead to adaptation, which is not the case, it largely depends on the available variation, which in turn depends on changes at the molecular level, called mutations. If there is no variation, there is no selection, and the available variation is the outcome of stochastic molecular changes. One might then argue that evolution is, in a sense, not selection-driven but mutation-driven. It is the available genetic, and the resulting phenotypic, variation that constrains or allows evolution to proceed in various directions. And this variation depends on stochastic molecular processes.

Mutations, or changes in DNA sequences, can produce new characteristics. These include changes within DNA (coding or regulatory) sequences or larger rearrangements in chromosomes (insertions, duplications, deletions, etc.). All of these are stochastic events, as chance plays a major role in their occurrence: Whether or not they will occur, and what their outcome will be is a matter of chance, in the sense that it cannot be predetermined or predicted.

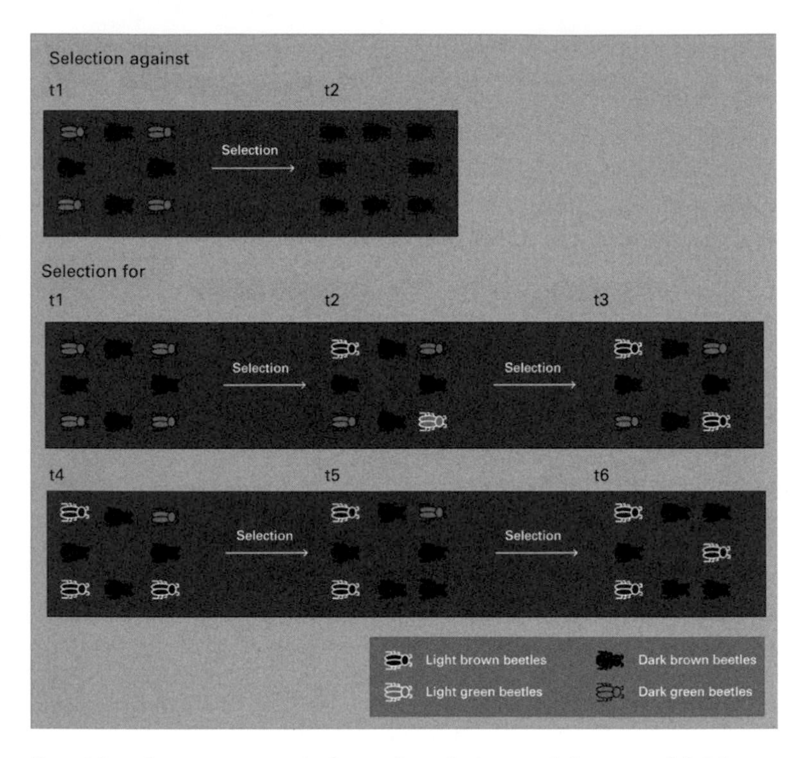

Figure 6.3 Selection *against* and selection *for*. In both cases dark green and dark brown beetles initially live in a dark brown environment. In the first case, only the dark brown beetles survive while the dark green ones are quickly eliminated; there is *selection against* the latter. In the second case, there is a gradual *selection for* brown color over many generations with a gradual accumulation of useful variations: not only dark brown beetles but also light brown ones can be adapted (in all cases, proportions, not actual numbers of the various types of individuals, are depicted).

For example, whether a nucleotide change in a DNA sequence will lead to the production of the same amino acid in the respective protein (silent mutation), or a different amino acid that might result in a different or a defective protein, is in large part a matter of chance. Similarly, whether unequal crossing-over will result in the duplication of a DNA sequence without harmful effects or in the disruption of a DNA sequence that is implicated in the production of an important protein is also a matter of chance. All changes at the molecular level are stochastic events because we cannot tell in advance when they will take place or what their outcome will be.

Stochastic events also take place at the organismal level. One very important case is that of horizontal DNA transfer, already discussed in Chapter 5. Some bacteria may have drug resistance due to a specific DNA sequence. Other bacteria may not have this property. However, it is possible that two bacteria are connected and that a plasmid is transferred from one to the other. If this happens, then the bacterium that receives the plasmid will also thereafter exhibit drug resistance. Whether such a DNA transfer from one bacterium to another will happen or not is a matter of chance, or in other words cannot be determined in advance. In this sense, this is a stochastic event.

Finally, stochastic events take place at the population level. The exemplar process of this kind is genetic drift or simply drift. Drift can be defined as a process of indiscriminate (parent or gamete) sampling. It can be better understood if it is contrasted to selection. Selection can be described as a process of discriminate sampling; some characteristics confer an advantage to their possessors, which in turn survive and reproduce better than other individuals that do not have these characteristics. Thus, there is selection for these characteristics and not for others, and in this sense (gamete or parent) sampling is discriminate (some characteristics are sampled through the reproduction of their possessors but not some others). In this case, differences in reproductive success are due to differences in corresponding characteristics that facilitate the survival and reproduction of their bearers. In contrast, drift is a process of indiscriminate sampling. In this case, reproductive success is not due to selection but depends on which characteristics happen to be sampled, which in turn is a matter of chance.

Imagine a forest in which green and brown beetles live and where there is no selection for color (Figure 6.4). This population consists of four different varieties of beetle (two varieties of green beetles – dark green and light green – and two varieties of brown beetles – dark brown and light brown). As there is no selection for color in the particular environment, each variety constitutes 25 percent of the beetle population. Assuming that frequencies are stable because there is no selection, up to t2 there is no change. However, a forest fire happens to kill more green beetles than brown ones. As a result, the

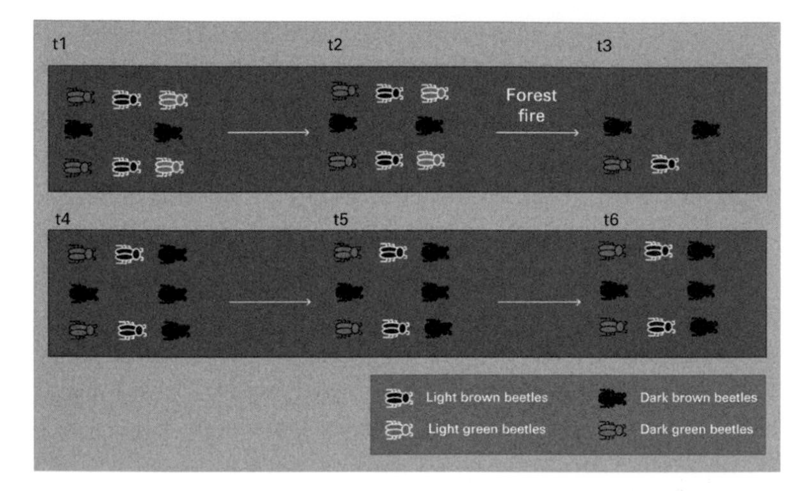

Figure 6.4 Drift due to indiscriminate parent sampling. Only those individuals that happened to survive produced offspring (in all cases, proportions, not actual numbers of the various types of individuals, are depicted).

brown varieties are now 75 percent of the remaining population, whereas the green ones are just 25 percent. When the population reaches the size before the fire, the frequencies are still the same and because there is no selection for color they might remain the same for some time. In this case, the proportion of each variety in the population changed because some individuals but not others happened to die before reproducing. There was a parent sampling (some beetles had offspring, others did not), but it was indiscriminate. The frequency of brown-colored beetles increased in the population without any selection *for* this characteristic.

However, not all stochastic processes are as undirected as drift. One interesting case is genetic draft. Draft is another name for linked selection, already discussed in the previous section. Imagine that a DNA sequence G_{A1} is implicated in brown coloration in beetles and it is linked to a DNA sequence G_{B1} that is implicated in the black coloration of the internal part of their wing blades. The respective alleles are G_{A2} (green coloration) and G_{B2} (white wing blades). In an environment where brown coloration confers an advantage to beetles, compared to green coloration, there will be selection *for* brown color. However, there will also be selection *of* black wing blades as a result of the process of genetic draft. Black wing blades will become the prevalent characteristic in the population, not because there was selection for them, but only because the DNA sequence G_{B1} happened to be linked to the DNA sequence G_{A1}.

These are some examples of stochastic events (mutation, horizontal DNA transfer) and processes (drift, draft) that can drive evolution in one or another direction unpredictably. A usual criticism against evolution is its "randomness"; critics state that it is impossible for complex systems to occur through random events. However, evolution is not a random process; it is just that particular events and processes can have a high degree of unpredictability. For example, whereas one might predict the outcome of selection, it is not possible most of the time to predict the outcome of drift because one cannot know in advance which of the events or processes, such as those described above, will take place and when. Interestingly, diversity and complexity can arise by the simple accumulation of accidents and eventually increase on average. This can be illustrated with the example of a picket fence. Such a fence may consist of pickets which initially are identical to each other. However, as time goes by different things can happen to different pickets (a pollen grain stains one picket; a passing animal knocks a chip of paint off another; the bottom of another one becomes moldy and crumbles where it touches the ground, etc.). As a result, the pickets become different from each other and this process can continue indefinitely. Eventually, there is an increase in the complexity and the diversity of the fence as it consists of pickets that over time become very different from one another. In this sense, undirected, unpredictable, stochastic processes can have dramatic effects.

One important concept that helps describe the implications of stochastic processes and events is contingency, proposed by Stephen Jay Gould, who argued that the history of life is not predictable as organisms have evolved through a series of contingent events. He illustrated the idea of contingency by the metaphor of the tape: "You press the rewind button and, making sure you thoroughly erase everything that actually happened, go back to any time and place in the past ... Then let the tape run again and see if the repetition looks at all like the original"; "any replay of the tape would lead evolution down a pathway radically different from the road actually taken." In this view, the history of life on Earth has been determined by contingent events.

Contingency has two main features: unpredictability and causal dependence. There are several possible evolutionary paths (contingency); it is impossible to predict in advance which of these is going to actually be taken (unpredictability), but there are certain constraints in the possible outcomes once a specific pathway is taken (causal dependence).

Here is an example to illustrate this. Imagine a population consisting of equal numbers of different varieties of brown and green beetles, living in an environment where there is no selection for either color. How should one expect this population to evolve? It might remain as it is for years. However, if this population migrated to a brown environment (or if brown color somehow became the dominant one in their current environment – e.g., due to the destruction of vegetation) then this population might evolve to one consisting of brown beetles only (outcome B in Figure 6.5). Similarly, if this population migrated to a green environment (or if green color somehow became the dominant one in their current environment – e.g., due to an increase in vegetation) then this population might evolve to a population of green beetles only (outcome G in Figure 6.5). What is the most probable outcome? No one

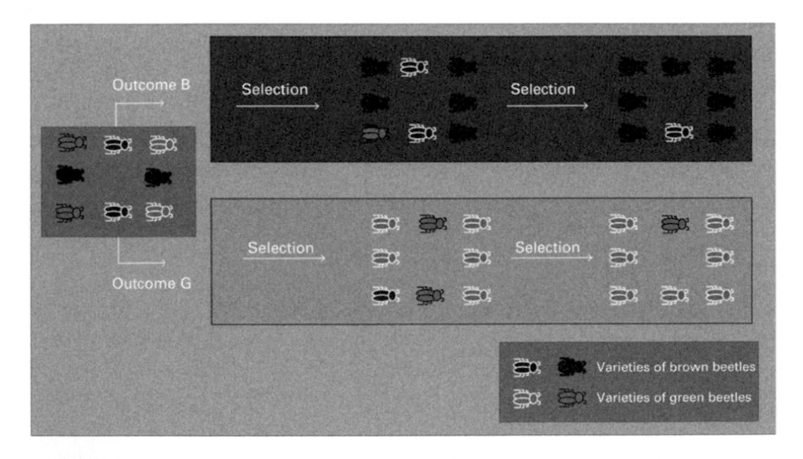

Figure 6.5 Contingency in evolution. Which of the two evolutionary paths will be taken is unpredictable (contingent *per se*). Once a path is taken there is a causal dependence of the outcome on the antecedent conditions (contingent *upon* it) (in all cases, proportions, not actual numbers of the various types of individuals, are depicted).

can tell in advance. Beetles might migrate or their environment might change, but this cannot be known in advance (unpredictability). Now, if one of these two evolutionary paths is taken, it will determine the outcome of evolution (B or G). Depending on the antecedent conditions, the initial population might evolve to one consisting of either varieties of green beetles only or of brown beetles only (causal dependence). Evolution thus depends on turning points – critical events that are contingent *per se*, and *upon* which future outcomes are also contingent. Whether outcome B or G will occur is unpredictable; in this sense outcomes B and G are contingent *per se*. Once B or G occurs, any future outcome is contingent upon the conditions that drove the evolution of the population to one or the other direction.

Speciation, Extinction, and Macroevolution

Many of the questions asked by evolutionary scientists are about the past. Understanding the answers that they provide requires overcoming an important conceptual issue: the perception of deep time, or the fact that evolution has taken place on Earth for a vast amount of time that is difficult for us to perceive. One way to consider time spans is to compare actual time to a 24-hour equivalent. Table 6.1 presents some major events in evolution in actual timing and in a 24-hour equivalent.

It is evident from Table 6.1 that some events that we may consider as quite old, such as the extinction of dinosaurs, are not that old after all: just 21 minutes ago if we consider that the Earth was formed 24 hours ago. It seems that for the first six hours there was no life on Earth. Prokaryotic cells were the only form of life for almost the next 12 hours (half of the age of Earth!), and eukaryotic cells appeared a bit more than six hours ago. Dinosaurs only appeared about an hour ago and they disappeared just 21 minutes ago. Our lineage diverged from that of our closest relatives (chimpanzees) less than two minutes ago, and we have been here in our current form for much less time than it took you to read this paragraph. These are important to keep in mind when thinking about evolution. The amount of time available for evolution is immense, and not easy for us to perceive.

The dates in Table 6.1 are estimated by dating the rocks in which fossils are found. This is done by using the principles of radioactivity in what is called radiometric dating. Simply put, this technique is based on the comparison of

Evolutionary event	Actual time (million years ago)	24-hour equivalent (hours, minutes, seconds ago)
Formation of Earth	4567	24 hours
First prokaryotic cells	3400	17 hours, 52 minutes
Accumulation of oxygen in the atmosphere	2400	12 hours, 37 minutes
First eukaryotic cells	1200	6 hours, 18 minutes
First animals	570	3 hours
First plants	470	2 hours, 28 minutes
First invertebrates	420	2 hours, 12 minutes
First vertebrates	380	2 hours
Dinosaurs	225	1 hour, 11 minutes
End-Cretaceous extinction	66	21 minutes
Split of chimpanzee and human lineages	6	1 minute, 53 seconds
First members of genus *Homo*	2.8	52 seconds
Modern humans	0.2	3.8 seconds

Dates from Knoll, A. H., & Nowak, M. A. (2017). The timetable of evolution. *Science Advances*, 3(5), e1603076.

Table 6.1 Some major events in evolution in actual time and in a 24-hour equivalent

the measured abundance of a naturally occurring radioactive isotope and its decay products, as well as its half-life, i.e., the time required for the isotope to decay to half of its initial quantity. Isotopes are different forms of the same chemical element with slightly different masses, because the nuclei have the same number of protons but different numbers of neutrons. For example, the radioactive isotope of carbon-12 is carbon-14; when organisms die, carbon-14 decay begins and so the quantity of this isotope decreases exponentially. The half-life of carbon-14, i.e., the time required for half of its quantity to decay away, is 5730 years. Thus, by measuring the remaining amount of carbon-14 in some material, one is able to estimate its age. Other isotopes, such as uranium-238 with a half-life of 4.7 billion years, can be used for this purpose.

An important evolutionary phenomenon is speciation, or the emergence of a new species. But how does it occur? How do two populations that originally belonged to the same species come to be reproductively isolated from each other so that they can be regarded as distinct species? Roughly put, there are two main processes of speciation at the two extremes, and a continuum between them. On the one hand, speciation can take place when two populations are geographically isolated from each other, e.g., because of a mountain or a river between the areas they live in. This process is called *allopatric speciation*. In this case, individuals of the two populations do not meet at all, and any new genetic variants are restricted to the population in which they occur and cannot be passed to the other. As a result, the two populations may evolve independently and eventually diverge from each other. The other form of speciation can take place when a population evolves to two or more reproductively isolated groups that are not geographically isolated. This is the process of *sympatric speciation*, in which individuals encounter each other and are able to reproduce while they diverge. Between these two extremes one can find cases of *parapatric speciation*, a process through which distinct species evolve from populations that are somehow, but not completely, isolated geographically.

Figure 6.6 provides examples of different kinds of situations that can lead to speciation. In the case of allopatry, the two populations are kept apart from each other because of a geographic barrier. In the case of parapatry, the two populations interact at the edges of their habitats. In the case of mosaic sympatry, individuals from the two populations live together but in particular parts of the area they inhabit. Finally, in the case of pure sympatry, the individuals from the two populations live together throughout the whole area. What the figure shows is that a continuum from allopatry to pure sympatry can exist. Allopatry is the major condition leading to speciation, and so allopatric

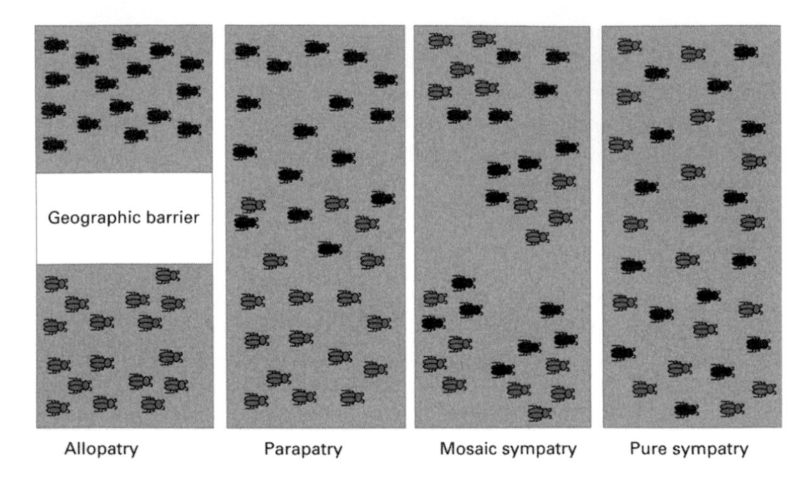

Figure 6.6 The continuum between allopatry and sympatry.

speciation is perhaps the most usual process. Under particular circumstances, allopatric speciation can lead to extensive diversification, as in the cases of adaptive radiation, when new species emerge from an initial one and adapt to previously unoccupied niches. The traditional example here is that of Galápagos finches, but there are other cases like this, such as *Anolis* lizards.

Whether in the same or different habitats, what is important is that significant diversification of populations and divergence from an initial state is possible. The environmental conditions under which speciation can take place are important, but what is more crucial to understand is how organisms diversify. Environmental conditions can cause selection for a particular characteristic; different kinds of barriers causing reproductive isolation may promote divergence of two different populations to two distinct species. Divergence results from the accumulation of different genetic variants in either of the two populations because their individuals do not mate and so do not give rise to offspring with shared DNA sequences. Consequently, and depending on the genetic changes that will take place in the course of evolution, two populations may diverge significantly so as to end up being reproductively isolated. Allopatric speciation gives the clearest example of how this can happen, since geographically isolated populations do not mate at all (Figure 6.6 provides a simple illustration of how allopatric speciation might be initiated). In contrast, sympatric populations may not diverge significantly because they mate and produce offspring that have several combinations of DNA sequences that are shared by several members of the two populations, although it has been shown that sympatric speciation is indeed possible.

But how does this divergence take place? Once again, evolutionary developmental biology provides important insights, although further research is necessary. Studies show that changes in developmental processes can be implicated in speciation events. For example, it has been found that DNA sequences involved in developmental signaling and regulation are significantly more likely to be evolutionarily retained in multiple copies after duplication than are other DNA sequences, suggesting a role for developmental regulation in speciation. It has also been shown that amphibian and fish clades in which polyphenism – a form of phenotypic plasticity in which two or more distinct phenotypes are produced in different environments by the same genotype – has evolved are more species-rich than closely related clades without polyphenism. Another example is phenological isolation, which is isolation due to differences in the time to maturity or reproductive activity, which is one of the outcomes of changes in the timing of developmental processes. Therefore, the morphological changes caused by changes in developmental processes may be implicated in speciation because they produce significant reproductive barriers.

This brings us to an important conclusion: In the continuum from allopatry to pure sympatry, different kinds of barriers may cause populations of the same species to diverge significantly and become reproductively isolated. However, in many cases it is difficult to set strict limits on when this happens. A clear example of this is the so-called ring species, which are considered to illustrate stages in the process of speciation because they include a full array of intermediate conditions between well-marked species and geographically variable populations. In such cases, neighboring populations can interbreed successfully, whereas those that are geographically more distant cannot (Figure 6.7). This shows that the process of speciation should be better perceived as constituting a continuum, too.

Here is then what may happen. Two populations sharing common ancestry may initially diverge, evolving different characteristics (anagenesis). Divergence

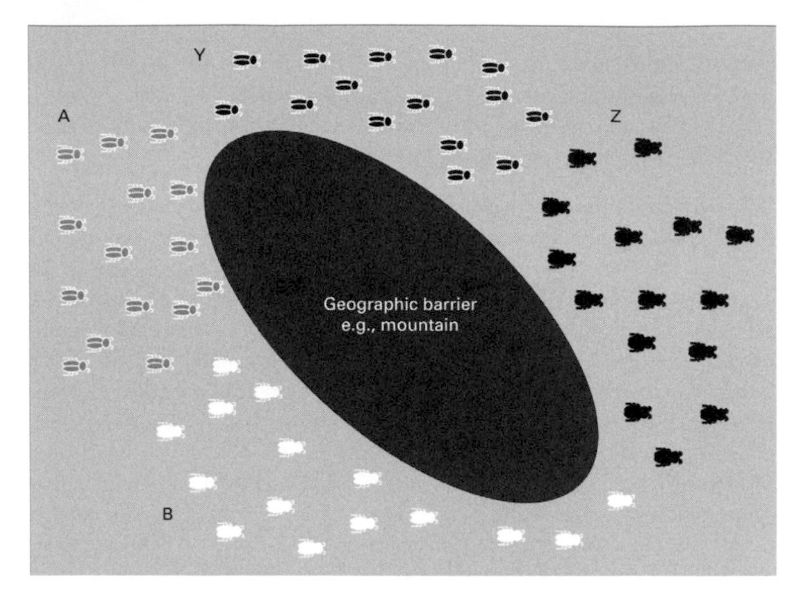

Figure 6.7 An illustration of a so-called ring species. Populations A and Y initially evolved from a common ancestral population. Individuals from populations living close to each other can interbreed successfully (B with A; A with Y; Y with Z), whereas those living further away cannot. Although populations B and Z live close to each other, they cannot interbreed successfully because they have diverged and currently are reproductively isolated. In this case, B and Z can be considered as distinct species. This could also be the case for A and Z, as well as for B and Y (in all cases, proportions, not actual numbers of the various types of individuals, are depicted).

may continue, or long periods without significant change (stasis) may occur. Such a divergence may eventually give rise to new species and the initial lineage may split into two or more lineages (cladogenesis). Throughout this process, extinction is always a possibility. Hence, the important point in the study of speciation is to understand in which part of this continuum a species is actually found. One problem is not so much the incompleteness of the fossil record, but the fact that one cannot deal with "species" in the same way when dealing with extant and extinct organisms. In particular, in the case of lineages for which considerable anagenesis but no cladogenesis occurs, specimen samples, e.g., at 10 million-year intervals, can be so

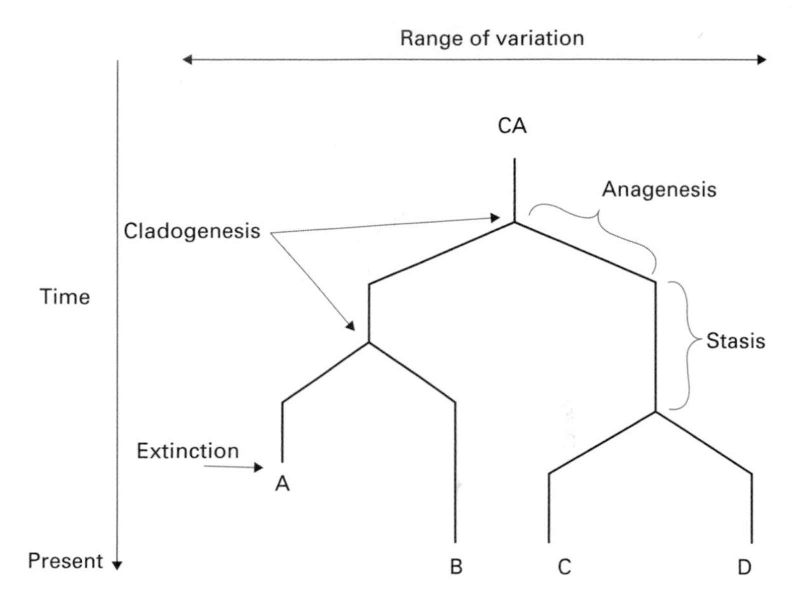

Figure 6.8 The continuum of speciation. Species may split to more lineages, and evolve by diverging, remaining the same, or going extinct. (CA: common ancestor; A–D: species).

different that they can be considered as separate species. However, these are not comparable to extant, distinct species derived from repeated cladogenesis from a common ancestor. Figure 6.8 provides an illustration of this continuum.

Extinction may be perceived as an exceptional case, but it is actually the rule in evolution. It seems that the vast majority of all species that have ever lived on Earth have gone extinct. Although extinction events can take place here and there, of considerable interest are those massive events described as mass extinctions. Such events can have a profound impact. One such example is the end-Cretaceous or K–Pg (K is the abbreviation for the Cretaceous period and the Tertiary period has been divided up into Paleogene and Neogene periods) mass extinction, which took place around 66 million years ago and famously caused the extinction of non-avian dinosaurs. This mass extinction largely determined the taxonomic and biogeographic characteristics of today's organisms. Extinction events can have a long-lasting impact on biodiversity and consequently on evolution. For example, the extinction of dinosaurs made possible the evolution and diversification of mammals, among other taxa, something that was likely not possible before the extinction due to predation on mammals. Available data suggest that clades with a wide geographic range are more resistant to extinction than other clades with narrow ranges. A probable explanation is that perturbations operate at a local scale so that those clades with a wide geographic range are less affected. However, this correlation changes in the case of mass extinctions, as events have a larger impact, and even clades with a wide geographic range are affected.

Speciation and extinction are included among those phenomena usually described as macroevolutionary. The distinction between micro- and macroevolution is useful because there are major differences between them. Microevolution encompasses phenomena of evolution within a species, whereas macroevolution encompasses phenomena across species. The important difference is that we can observe microevolutionary phenomena because in many cases (especially in microbes) they occur within a short time span. But such changes are also observable in multicellular organisms such as Galápagos finches. In contrast, it is almost impossible to observe macroevolutionary phenomena because they are usually completed over very long time frames. Distinguishing between these two is important because they are often confused. Thus, unanswered questions about the latter are sometimes deliberately used by antievolutionists to question the foundations of the former. But this is entirely wrong. The fact that we do not know all the details about a macroevolutionary process does not affect our understanding of the main processes of microevolution, such as natural selection or drift, for two reasons: (1) because microevolutionary processes can be (and actually have been) demonstrated in the lab or in the wild; and (2) because we may eventually come to know more about macroevolution.

One important conclusion is that, other phenomena notwithstanding, selection has a significant role in macroevolution. As I have already described, selection for some characteristics can drive the evolution of a population and eventually produce changes in its genetic and phenotypic structure. This is selection *within* the species level. However, there can also be selection *at* the species level, described as species selection. There are two senses of species selection: (1) a broad sense according to which speciation and extinction depend on characteristics at the organism level, such as body size and fecundity; and (2) a strict sense according to which speciation and extinction depend on characteristics that are emergent at the species level, such as geographic range and population size. It is important to note that these characteristics may affect speciation and extinction in different ways and so should be studied carefully before conclusions are reached. In order to understand speciation and extinction, we need to consider all these different characteristics, their possible interactions, and their effects.

But how does macroevolution proceed? Figure 6.9 presents all possible combinations of the variation in rate of evolution between different lineages over evolutionary time (tempo) and the mechanisms driving these varying rates of change (mode). All these combinations have been recorded in the fossil record. The mode of evolution includes anagenesis and cladogenesis, already discussed previously. Anagenesis is the evolutionary divergence of a

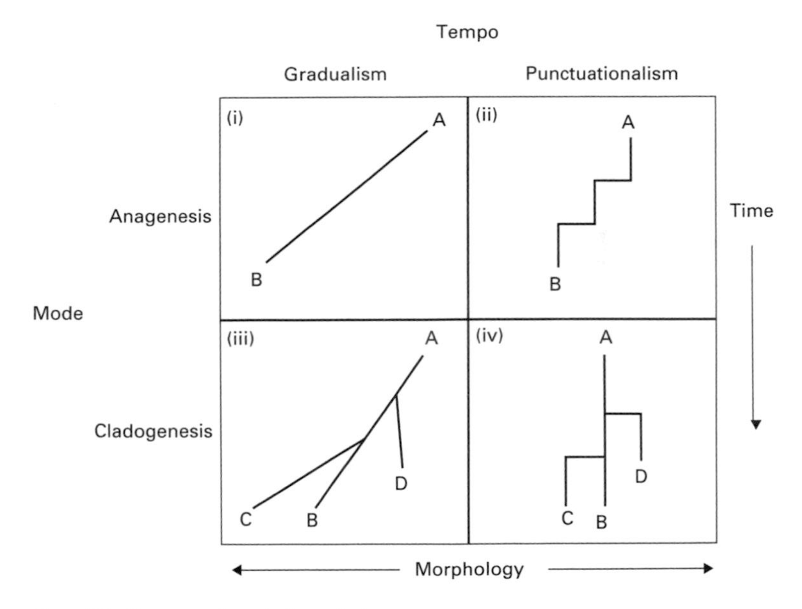

Figure 6.9 Tempo and mode in evolution.

lineage over time, whereas cladogenesis is the splitting of a lineage into two or more. The tempo of evolution includes gradualism and punctualism. The main difference between them is that in the first case all intermediate forms are found in the fossil record, whereas in the second they are not. According to the idea of punctuated equilibrium, the evolutionary histories of most species display stasis, absence of significant change, which is punctuated by rapid morphological evolution associated with cladogenesis. This model of punctuated equilibrium was offered as an alternative to the slow, continuous evolution of gradualism.

Interpreting macroevolutionary patterns and explaining macroevolutionary processes is conceptually challenging. Evolution within the species level can be understood more easily, because studies of natural populations and laboratory experiments are possible. In contrast, macroevolutionary processes require access to the deep past, which is difficult to achieve. The challenge is to observe the extant morphological diversity and combine such observation with fossil data in order to explain its evolution. The important component in this case is time. This is what makes evolutionary explanations, and particularly macroevolutionary explanations, distinctively historical.

Evolutionary Explanations and the Historicity of Nature

Evolutionary scientists use the available data to make inferences to the best explanation (IBE). The central idea is that explanatory considerations are a guide to inference: Scientists make an inference from the available data to a hypothesis that would, if correct, best explain the data. According to IBE, hypotheses are supported by the available data they are supposed to explain; data support the hypotheses precisely because they could explain it, and it is only by asking how well various alternative hypotheses could explain the available data that one can determine which hypotheses should be accepted. An important distinction is that between potential and actual explanations. A potential explanation is one that satisfies all the conditions of actual explanation, with the possible exception of truth. According to IBE, we infer that what would best explain the available data is likely to be true. Thus, the best potential explanation is likely to be an actual explanation. There are two important advantages of IBE: (1) it is context-dependent, i.e., a particular scientific hypothesis would, if true, explain particular observations; and

(2) it discriminates between different hypotheses, all of which would explain the data, since it points to the one that would best explain it.

Based on these considerations, IBE could be defined as inference to the best of the available competing explanations, when the best one is sufficiently good. But how good is "sufficiently good"? Does this refer to the most probable explanation or to the explanation that would, if correct, provide the greatest degree of understanding? Let me clarify the difference with an example. We know that HIV causes death because of opportunistic infections. One explanation is that HIV causes deficiency of the immune system. This explanation is likely but provides no understanding. Another explanation is that HIV destroys T-cells. This is also likely and provides understanding, and as such it is the best one. Thus, the explanation that would, if correct, provide the most understanding is the explanation that is judged as the most likely to be correct. To do this, we should contrast it to other alternative explanations. Thus, we should not simply ask "Why A?" but rather "Why A rather than B, C, etc.?" In this way, what would count as the best explanation would depend on both A and B, C, etc., and would identify a cause that made the difference between A and B, C, etc.

Let us consider the K–Pg mass extinction that took place about 66 million years ago and is famously the one responsible for the extinction of non-avian dinosaurs. Before 1980, many competing explanations had been proposed by paleontologists for the K–Pg extinction, including changes in oceanographic, atmospheric, or climatic conditions, a magnetic reversal, a nearby supernova, volcanism, or the flooding of the ocean surface by water from a postulated Arctic lake. However, the available data did not provide strong support for any of these alternative hypotheses. It was thus assumed that the so-called K–Pg boundary might provide important information about this event. The K–Pg boundary is a 1 cm thick, distinct, thin layer of clay between two layers of limestone that are chemically similar to each other. The K–Pg boundary is found all over the world and marks the end of the Cretaceous period and the beginning of the Paleogene period. Walter Alvarez, a geologist, and his father Luis, a physicist, used the element iridium as a clock because it can be measured at low levels and because it mostly comes from meteoritic dust. They found that clays from the K–Pg boundary contained iridium levels more than 30 times higher than the limestones on either side, which was too much to be explained in terms of known geological processes. Later studies confirmed the presence of an iridium anomaly in the K–Pg boundary more than 100 times higher compared to the background.

The Earth's crust does not contain much iridium because it is a heavy element and most of it sank into the mantle and core during the formation of the planet. Although not all meteorites are rich in iridium, asteroids and comets from the formation of the Solar System usually have higher concentrations. However, volcanism also brings mantle material to the surface and so this was a plausible explanation for the iridium anomalies. Meteorite impact and volcanism thus became the only two alternative explanations for the iridium anomaly, because none of the other competing hypotheses could explain it. Further research supported the meteorite impact over volcanism. Analysis of K–Pg boundary sediments showed large quantities of mineral grain, predominately quartz, exhibiting a highly unusual pattern of fractures. Sudden application of extremely high pressure is required to fracture minerals in this way. The observed mineralogical features were characteristic of shock metamorphism and were considered as evidence that the shocked grains were the product of a high-velocity impact between a large extraterrestrial body and Earth. Further studies showed that lamellar deformation features in quartz from tectonic and explosive volcanic environments only superficially resemble features from known shock and/or impact environments.

The excess iridium and shocked quartz in the K–Pg boundary were evidence suggesting that a huge meteorite hit Earth 66 million years ago. However, this did not necessarily suggest that the mass extinctions were caused by the meteorite impact. In an analysis of fossil record data (accumulated over 12 years from seven measured sections in France and Spain) of end-Cretaceous macroinvertebrates, 40 molluscan species were recovered. From these, only two seem to have survived into the Paleogene and three were excluded from the analysis because they are each known from a single fossil. Thus, the fossil records of the 28 species of ammonites and seven species of inoceramid bivalves were analyzed. Of these, six ammonite species appear to have become extinct before the K–Pg impact event, most likely due to background extinction processes. All but one of the inoceramid bivalve species became extinct well before the K–Pg impact event, perhaps due to global changes in deep-sea circulation. All the remaining species (22 ammonites and 1 inoceramid bivalve) seemed to have been possible victims of the K–Pg mass extinction. The point made here is not that the meteorite impact event is not the main cause of the mass extinction, but that we do not (and perhaps cannot) know all the details. The meteorite certainly had an impact, but other processes may also have contributed to the K–Pg mass extinction. Indeed, more recently some paleontologists have developed a new volcanism hypothesis, arguing that massive volcanic eruptions in the Deccan plateau region of India, which seem to have released large quantities of toxic gases, could have contributed to the K–Pg extinction. Details notwithstanding, the explanation of the K–Pg extinction exhibits two important characteristics: that causes can be identified by their effects; and that it is important to obtain evidence from multiple independent sources.

Let us consider the first of these features with an example. Imagine you throw a ball at a window. Under what conditions will a window break when a ball is thrown at it? In this case, we can consider a moving ball as the cause and the broken window as the effect. The ball has some features with causal influence, i.e., due to which the effect takes place. Some of these are the weight of the ball, the material of which the ball is made, the size of the ball, and the speed with which the ball falls on the window; all of these have a causal influence on the effect. If one of these features is different, a different effect may occur. Several combinations of these features can have the same effect: a broken window. However, if one of them is different, the window may not break. For example, if a tennis ball is thrown at high speed at the window, the window will most likely break. But a table tennis ball thrown at high speed will not break the window. Thus, speed alone is not sufficient for the window to break; the material of which the ball is made matters, too. But neither is the material alone sufficient to break the window. A tennis ball thrown at low speed will probably not break the window, whereas a bowling ball likely will. This means that considering each of the features with the causal influence on its own is not enough to guarantee that their effects will occur. However, even if we do not know all the details, we can sufficiently explain why a window broke. If we observe dispersed pieces of glass and a ball on the floor, these are traces of an effect – in this case the breaking of the window. We do not need every piece of glass in order to explain how the window was broken. Nor do we need every detail of the causal history of the event, e.g., what the weight of the ball is, what exactly it is made of, what its speed was, etc. in order to conclude what happened. Thus, we can identify a cause by just observing traces of its effect, even if we are not aware of all the details.

The second feature of the explanation for the K–Pg extinction is the availability of independent lines of evidence for the meteorite impact. These include the existence of excess iridium and shocked quartz in a thin layer of sediment found all over the world, as well as the lack of particular ammonite fossils after that. Together, these are evidence for the impact of a huge meteorite as the best explanation for these observations. These independent observations provide the foundations both for the inference to the meteorite impact event and for selecting that one over other competing hypotheses, such as volcanism. Interestingly, the more recent argument in support of the impact of the Deccan volcanic eruptions also cites multiple lines of evidence, such as the intensification of these volcanic eruptions, and radical changes in the fossil record during the years before the meteorite impact.

At this point I should note that trying to identify the details of an event that had a significant impact is actually the attempt to describe one important component of evolutionary explanations: the antecedent conditions. For example, in explaining adaptations, a lot of emphasis has been given to the role of natural selection, perhaps overlooking the fact that natural selection takes place as long as some antecedent conditions exist. For instance, there will be no selection in a population that has no variation, because there is nothing to be "selected." It is antecedent conditions like variation in a particular environment that cause natural selection, which in turn brings about evolution and, perhaps, adaptations. And it is not only whether natural selection will take place, but also its direction that is affected by antecedent conditions (e.g., what kind of variation was available in a population, or in what kind of environment was this population living?).

To illustrate the importance of antecedent conditions, let me use an example (Figure 6.10). A particular population of beetles might evolve in different ways, through different processes, or not evolve at all, depending on the antecedent conditions. For example, a population consisting of 50 percent green and 50 percent brown beetles might: (1) evolve to a population consisting exclusively of brown beetles through natural selection if the brown beetles have an advantage in the particular environment; (2) evolve to a population consisting exclusively of green beetles if during a fire only green beetles happen to survive from the initial population, even if the environment would have otherwise favored the survival of brown beetles (drift); and (3) remain as it is for numerous generations, as long as green and brown beetles have equal

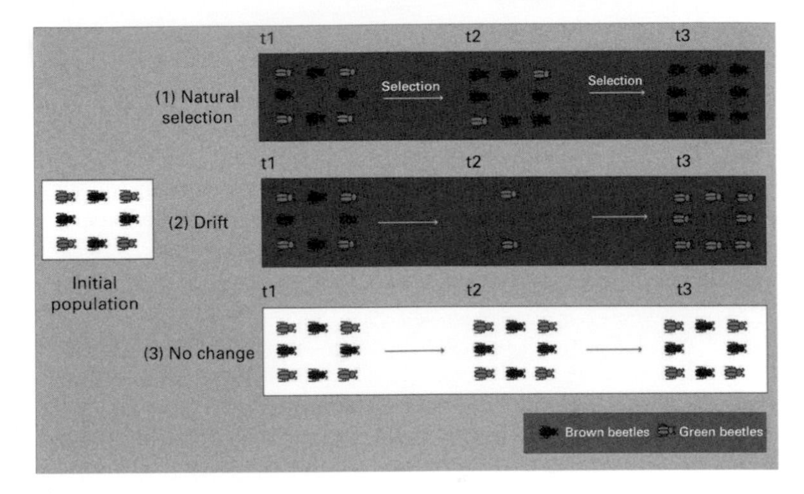

Figure 6.10 The importance of antecedent conditions in evolution: (1) natural selection, (2) drift, (3) no change (in all cases, proportions, not actual numbers of the various types of individuals, are depicted).

chances to arise through reproduction and neither of these two types has an advantage over the other in the particular environment. In all these cases, what is causally important is not only the process that does (or does not) take place, but also – and perhaps mostly – the antecedent conditions that cause each process or cause no process at all. In (1), the antecedent conditions are the variation within the population as well as the fact that brown color confers an advantage to its possessors in this particular environment (in another environment the outcome could have been different). In this case, it is the antecedent conditions that cause natural selection. In (2), the antecedent conditions are the variation within the population, as well as the fire that killed all brown beetles. If there is no fire, no such change might take place. In this case, it is the antecedent conditions that cause drift. Finally, in (3), the antecedent conditions do not cause any process of change but maintain the population structure. Although variation exists within the population, it is nevertheless not enough to cause evolution as long as green and brown beetles have equal chances to arise through reproduction, and neither of them has any adaptive advantage in the particular environment. As is obvious from Figure 6.10, different outcomes are possible: The population may evolve to one consisting exclusively of brown beetles, or green beetles, or not change at all. It is these elements of evolutionary explanations, the antecedent conditions and the subsequent evolutionary processes, which make these explanations historical.

7 Evolutionary Theory and the Nature of Science

Understanding the Nature of Science

The two central ideas of evolutionary theory are (1) that all organisms – both those currently living and those that have ever lived on Earth – are related through descent from common ancestors; and (2) that they have evolved or died out through natural processes. Simply put, evolutionary theory suggests that we are part of this world; we are one biological species among numerous others, with which we are more or less related, having diverged from our common ancestors over time. However, many people do not accept these ideas. They think that if we accept that we are animals, we are devalued and human morality is threatened. They also think that if we accept that life has no inherent purpose, this deprives it of meaning. For these people, evolutionary theory is a rather nihilistic idea, and this is where many of them see the conflict between evolutionary theory and religion. This conflict is also perceived by those proponents of evolutionary theory who are irreligious or atheists, and who also think that evolutionary theory has been the death blow of religion.

This seems like a straightforward situation to many. Evolutionary science shows that we are evolved animals; that there are no signs of intelligent design (ID) in this world; and that death is the guiding force of life. If you are pro-science and you accept this picture, you have to reject religion; if you are religious and unwilling to accept this picture, then you have to reject science. The conflict between science and religion seems inevitable, because the strict naturalism of science cannot coexist with or accommodate the supernaturalism that is inherent in religion. What is left, then, is the contrast between rationality and superstition, between evidence and belief, between science and religion – for those who are willing to see it, or so the argument goes.

However, I believe that the situation is a lot more complicated than this. It is not that I do not see a conflict between evolution and religion; I actually do. But that I could provide you with a rational, normative argument about their incompatibility does not mean that I am correct. Why? Because there is ample evidence that many religious people accept evolution, as well as that many evolutionary scientists are religious. Can I ignore this evidence simply because it contradicts my own views about the incompatibility? No, and this is why the distinction between knowing and believing made in Chapter 2 is crucial. The point here is that those of us who are proponents of science should not ignore the evidence about the compatibility of evolution and religion – that is, that there are people who accept both. Does this entail that science and religion are compatible? No, not all. That people believe in compatibility does not mean they are right. My point is that the situation is far from a simple dichotomy, and that several philosophical – not scientific! – arguments about the relationship between science and religion are available.

Beyond the question about the conflict, there is a broader issue that I consider as more important: how people understand the nature of science, or what science is and what it can achieve. This is often reflected in the arguments of antievolutionists, who attack evolutionary theory and try to devalue it in various ways. For instance, a claim antievolutionists often make is that they do not accept evolutionary theory because it is "only a theory." What they fail to understand is that the word "theory" has an entirely different meaning in science compared to its colloquial use. In the latter case the word "theory" is used to denote a hypothesis, a thought, or a speculation. In our attempt to explain an event we might say "I have a theory," and this means that we are simply trying to provide an explanation for this event based on what we happen to know at that time. This is very different from the meaning of the word "theory" in science. A scientific theory is a collection of models, principles, assumptions, and ideas that can provide explanations and predictions about natural phenomena and relate to an area of inquiry with widely accepted methods and foundations and a strong empirical basis. Evolutionary theory is not an assumption or a hypothesis; it is a solid scientific theory.

Another argument of antievolutionists is that there are several phenomena that evolutionary theory cannot explain. Therefore, the argument goes, evolutionary theory is not credible and should be rejected. For them, something more than natural processes is required to explain the origin of species and their characteristics. This additional factor is God, whose existence and power are the ultimate explanation for everything. In this view, whenever there is a "gap" in the explanatory potential of science, God's divine intervention stands as the always-sufficient alternative, an approach that has been characterized as the "God of the gaps." However, the argument that "there is no scientific explanation for *X*, therefore *X* can only be explained by assuming divine intervention" is ambiguous. There is an important difference between the propositions "*X* is *unexplained* by science," which means it has not been explained by science *yet*, and "*X* is *unexplainable* by science," which means that science cannot *in principle* explain *X*. The former proposition refers to questions that science has not answered *yet*, but that it can in principle answer. That there is no scientific explanation for a phenomenon because, for example, no relevant evidence has been found *yet*, does not entail that such evidence will never be found.

A last, significant problem of antievolutionists is that they do not really try to understand how organisms have arrived at their present forms, but rather try to develop an account for this that agrees with religious texts. Whereas scientists study the evidence and try to reach conclusions, antievolutionists usually have the conclusions ready (and these vary among the different antievolutionist groups, from establishing the existence of God to establishing agreement with a literal reading of the Bible) and try to make the evidence fit with those. In other words, antievolutionists are arguing against evolution because it does not conform well with their worldviews. They are not open to revision because they are unwilling to change their conclusions. This is most evident in their unwillingness to accept the evidence for the fact of evolution. In contrast, scientists are relatively open to revising their understanding of the evolution of life on Earth. Unlikely as it might be for evolutionary theory to be rejected, no serious scientist would deny this possibility in principle, because this has been the case for many successful scientific theories of the past.

From all of the above, we can see that antievolutionists misunderstand and misrepresent the nature of science: what a scientific theory is, what it can explain, how it relates to evidence, and whether it is open to revision. Therefore, it is important that they, and everyone, achieve a good understanding of the nature of science as well, and not only of evolution. Science relies on theories, which can – in principle – explain everything about the natural world – even though they may have not explained everything up to now – and which are open to revision. These are some features of science. Is this all? No, there is more. A big issue is that we cannot easily demarcate science from non-science, because the habits of mind and the epistemic practices of scientists share several similarities with non-scientific endeavors. Nevertheless, based on what has been presented in this book, I can propose eight basic features of science:

1. Science studies the *natural world*.
2. All scientific conclusions rely upon *empirical evidence*.
3. These conclusions must be confirmed by *independent testing*.
4. If several conclusions are supported by empirical evidence and by independent testing, then scientists can accept the one that they *infer* to provide a better *understanding*.
5. It is *collective understanding* of the natural world, and not truth in any absolute sense, which is the ultimate aim of science.
6. The collective understanding that science produces is summarized in the form of *scientific theories that rely on abstractions and idealizations (models)* of the natural world.
7. Scientific theories are *open to revision or rejection* in the light of new evidence and new inferences that might bring about new understanding.
8. Because of its reliance on empirical evidence, independent testing, and its openness to revision, science provides the most *objective* means of understanding the natural world.

As this book is about understanding evolution, and as I have argued that "understanding" is the ultimate aim of science, let us now look in some detail what this is about. For this purpose, I draw on the theory of scientific understanding developed by the philosopher of science Henk de Regt. According to this theory, there are two basic values of science. First, scientific theories should agree with the observable world, a property described as empirical adequacy. Second, scientific theories should be internally consistent and not contain contradictory elements. These two values are prerequisites for a genuine understanding. Now, for scientists to understand a theory, it should be intelligible to them. Intelligibility is the value that scientists attribute to the qualities of a theory that facilitate its use. This is not an intrinsic property of a theory, but an extrinsic one because it depends on the skills of scientists who work with it. Theories are not intelligible or unintelligible *per se*, but intelligible or unintelligible to a particular group of scientists. In this sense, scientific understanding is the achievement of a community, as the work of different scientists may contribute to it. But what do scientists understand? It is important to distinguish between two types of understanding: (1) understanding a theory – that is, being able to use it; and (2) understanding a phenomenon – that is, having an adequate explanation of it. Obviously, (1) is a prerequisite for (2), because only intelligible theories can be used by scientists to construct models through which they can arrive at explanations of phenomena. Based on all this, de Regt has proposed the following criterion for understanding phenomena: A phenomenon is understood scientifically only if there is an explanation of it that is based on an intelligible theory and that exhibits empirical adequacy and internal consistency. It is important to note that having an intelligible theory is a necessary but not sufficient condition for scientific understanding. In order for an intelligible theory to produce understanding of phenomena, scientists must use it and arrive at an explanation that exhibits empirical adequacy and internal consistency. This is the case for the explanations provided by evolutionary theory, which has several other virtues that I present in the next section.

The Virtues of Evolutionary Theory

What constitutes a good scientific theory? This is a difficult question to answer, but philosopher Ernan McMullin has provided a useful list of the virtues of a good scientific theory: (1) empirical fit (support by data); (2) internal consistency (no contradictions); (3) internal coherence (no additional assumptions); (4) simplicity (testability and applicability); (5) external consistency (consonance with other theories); (6) optimality (comparative success over other theories); (7) fertility (novel predictions, anomalies, change); (8) consilience (unification); and (9) durability (survival over tests). A final virtue is explanatory power, which is actually a consequence of all the other virtues. Evolutionary theory is the theory that best explains the unity and the diversity of life on Earth. Let us now apply the above scheme to evolutionary theory in order to see why it is a very good scientific theory.

Empirical fit or support by data is a prerequisite in order for any theory to be considered as scientific. The main propositions of evolutionary theory, that all organisms on Earth share a common ancestry and that they all have evolved from preexisting ones through natural processes, are supported by all the available data from such diverse disciplines as paleontology, biogeography, molecular biology, cellular biology, genomics, and developmental biology. In Chapter 5 I explained that organisms very different in terms of structure and function share fundamental similarities at the cellular and the molecular levels. These similarities are easily explained with common ancestry: If two taxa T1 and T2 have evolved from the same ancestor A, it is anticipated that some of the characteristics of A will be found in T1 and T2. Differences are also explained by assuming that T1 and T2 evolved from A by divergence through natural processes such as natural selection or drift, described in Chapter 6. Over millions of years, evolutionary processes can produce very different life forms that nevertheless share crucial similarities. Studies of the development, the genomes, the geographical distribution, and the ecologies of contemporary species, as well as of fossils when they are available, point to the conclusion that evolution has occurred as described by evolutionary theory. The fact that we do not know some details yet, as well as that we may never know all the details, does not undermine how strongly evolutionary theory is supported by empirical data.

What is more important is that evolutionary theory exhibits internal consistency and no contradictions among its propositions. All data accumulated from different fields of research consistently point to the same conclusions. Organisms living in neighboring areas are found to be more closely related to each other – genetically speaking – than others living in more remote areas, even if their environments are not similar – the finches in the Galápagos are such an example. Similar structures may evolve in otherwise very different organisms – remember the wings of bats and birds; fossils exhibit similarities with extant species; embryos of closely related organisms (e.g., vertebrates) are very similar; and so on. Figure 5.10 shows the similarities in terms of DNA sequences between (a) humans, chimpanzees, and gorillas and (b) chordates, arthropods, and nematodes. It is no surprise that, despite the difficulties identifying homoplasy and its effects, there are more similarities among taxa in group (a) than among taxa in group (b). This coincides with what one would expect from comparing the structures of these taxa, and it is sufficiently explained by the fact that the time of divergence of taxa in group (a) from their common ancestor is more recent than the respective time for taxa in group (b).

Another virtue of evolutionary theory is its internal coherence, meaning that no additional assumptions are required. The principles and models of evolutionary theory can account for all observed phenomena, although it is not always possible to explain everything. What is important is that no additional arguments are required. There is nothing about the living world that evolutionary theory cannot explain in principle that would therefore require additional arguments. Of course, that evolutionary theory *can* in principle explain everything about the living world does not entail that it *will* actually explain everything. But the inability to explain every single aspect of the living world has to do with the lack of data or with human perception, not with the theory itself. As was explained in Chapter 6, achieving epistemic access to the deep past is very difficult; only traces of past events are available and these may not always adequately represent the actual events. This is why, for instance, recent evidence may make scientists reconsider their explanations for the K–Pg meteorite impact. But evolutionary theory can still sufficiently explain the extinction of non-avian dinosaurs.

Simplicity is another important virtue of evolutionary theory, and it relates to its testability and applicability (in the sense that a simple theory is easy to apply and test). Evolutionary theory can be easily tested, such as by making predictions about the distributions of species and comparing these predictions to biogeographical data. The discovery of *Tiktaalik* in Canada is a fine example of this. In his personal account of the discovery, evolutionary scientist Neil Shubin wrote that: "Most people do not know that finding fossils is something we can often do with surprising precision and predictability ... Of course, we are not successful 100 percent of the time, but we strike it rich often enough to make things interesting." Shubin then described how he and his colleagues took into account previous discoveries and decided where to look for fossils of organisms that would be intermediate forms between fish and tetrapods. They had to find rocks of the right age, of a type in which fossils would have been preserved and exposed at the surface. They were aware that amphibian fossils had been recovered from rocks about 365 million years old and that fish fossils had been recovered from rocks about 385 million years old. Consequently, they decided they should look for transitional forms in rocks aged 365–385 million years old. In addition, knowing that sedimentary rocks usually preserve fossils, they had to look for rocks formed in oceans, lakes, or streams, ruling out volcanic and metamorphic rocks in which fish fossils would not likely be found. Finally, they wanted to find areas that were not inhabited and where fossils might be exposed on the surface of rocks.

Shubin and his colleagues concluded that the Canadian Arctic was of the right age, type, and exposure, as well as unknown to vertebrate paleontologists. It therefore fulfilled all their criteria. And it was there, at the Fram Formation in Nunavut Territory, Canada, where *Tiktaalik* was eventually found.

Evolutionary theory also exhibits external consistency, which means that it exhibits consonance with other theories, such as those of chemistry and physics. All propositions made about how DNA sequences change and eventually evolve do not violate any of the laws/principles of physics and chemistry. At more complex levels of organizations, no matter if it is about how cellular processes, organismal characteristics, or population properties evolve, whatever evolutionary theory entails is in accordance with, e.g., the conservation of energy, the entropy of systems, or how chemical reactions take place. None of the explanations provided by evolutionary theory contradicts what we know about how atoms and molecules, or other components of complex systems, interact.

Another interesting aspect of evolutionary theory is its consilience – the enormous potential for unification. Evolutionary theory brings together and explains different kinds of data (fossils, biogeography, morphology, DNA sequences), which become evidence for the common ancestry of all life on Earth, discussed in Chapter 5, and of its evolution through natural processes, discussed in Chapter 6. The fossil evidence of extinct species, the distribution of extant species, as well as their morphology and DNA sequences, can all be explained by evolutionary theory. Populations are more closely related – genetically, morphologically, ecologically – to those living in neighboring areas, no matter how different their habitat is, than to those living in distant areas even in identical environments. Darwin's study of the Galápagos finches (Chapter 4) or contemporary studies of "ring species" (Chapter 6) point consistently to the conclusion that two populations are more closely related to each other the closer they live and the more they interbreed. This is best explained as the outcome of evolution.

Another virtue of evolutionary theory is its fertility. Evolutionary theory has made novel predictions and it has also been modified in the light of anomalies. For some reason, some people perceive the potential of a scientific theory to change in the light of anomalies as a weakness; however, this is a real strength. The critical point is what kind of change the theory undergoes. All scientific theories depend on auxiliary hypotheses. It is these hypotheses that often change, not the core of the theory. The opposite would be problematic because if the core of a theory changed, the theory would no longer exist. Darwin's theory can be summarized in the phrase "descent with modification": New taxa evolve through the modification of older taxa and as a result all of them share a common descent. This is still the core proposition of contemporary evolutionary theory. Since the *Origin* was published, evolutionary scientists have obtained a better understanding of evolution and they have also been accumulating evidence that supports evolutionary theory. The theory has of course itself evolved, as new pieces of evidence came in, anomalies arose, and modifications to address them took place; however, its core is still fundamentally the same.

A relevant virtue is the durability of evolutionary theory – its ability to surpass all tests. The legend has it that J. B. S. Haldane, a major contributor to twentieth-century evolutionary theory, once said that he would give up his belief in evolution if someone found a fossil rabbit in the Precambrian. This meant that one should not expect to find a mammalian fossil in rocks a few hundred million years older than the common ancestors of all vertebrates. In other words, given our knowledge of how the various taxa have evolved, we should not find any of them in strata and rocks older than we would expect. The cartoon *The Flintstones* depicted humans and dinosaurs living together, something entirely impossible given what we know about the extinction of dinosaurs 66 million years ago and the evolution of our species around 200,000 years ago (see Table 6.1). Finding the fossil skeleton of Fred Flintstone next to the fossil skeleton of Dino, the family's pet dinosaur, would be a real problem for evolutionary theory.

Another virtue of evolutionary theory is its optimality – its comparative success over other theories. Indeed, evolutionary theory offers the best explanations available for the characteristics of organisms by relying exclusively on natural processes. It is so successful that there is currently no other scientific theory about the origin of species and their features that scientists seriously consider. Evolutionary theory provides the simplest, most coherent, and most unifying explanations for what we observe. On the one hand, all similarities between organisms can be explained with reference to shared DNA sequences, a consequence of common descent (homologies) or of the evolution under similar conditions (homoplasies) (see Chapter 5). On the other hand, all differences observed between organisms can be explained with reference to different DNA sequences, a consequence of evolutionary processes such as natural selection and drift that have caused divergence of subgroups of what was initially the same population (Chapter 6).

Perhaps the greatest virtue of evolutionary theory, which is actually a consequence of all the other virtues discussed above, is its enormous explanatory power. This does not mean that evolution can explain everything about life and organisms. We do not always have all the required evidence to answer all kinds of questions related to the processes that give rise to species and their distinctive features or the patterns produced by these processes in DNA, fossils, and biogeography. But why, then, is evolutionary theory an explanatorily powerful one? The answer to this question is that evolutionary theory can explain a wide variety of phenomena based on a small number of propositions and models. It can explain the peculiar features and properties of organisms; the vast variety of life forms on Earth; the similarities of the DNA sequences of organisms which are morphologically very different; and much more. As already mentioned several times in this book, evolutionary theory can account for both the unity and the diversity of life on Earth. But we should be careful not to overextend the explanatory potential of evolutionary theory, as I explain in the next section.

Scientism and the Limits of Science

Some proponents of evolution have argued that the explanatory scope of science is not limited to the realm of the natural world, and that science is the only way of knowing in general. This attitude is described as *scientism*. Philosopher of science Susan Haack has noted that there is no simple way to determine whether the line between an appropriate respect for the achievements of science and an inappropriate deference to it has been crossed. However, she has provided a list of what she considers as characteristic indicators of this crossing:

1. *Forgetting fallibility*: Forgetting that science is an ongoing enterprise that makes claims many of which will not survive the test of time, and being ready to accept anything and everything bearing the label "science," or "scientific."
2. *Sanctifying "science"*: Using the terms "science" and "scientific" as synonyms for "strong, reliable, good."

3. *Fortifying the frontiers*: Insisting on a sharp line of demarcation between science and non-science.
4. *Mythologizing "method"*: Considering that what is distinctive about science is its method, which is supposed to be unique and distinct from other forms of inquiry.
5. *Dressing up dreck*: Adopting the tools of science to disguise a lack of real rigor.
6. *Colonizing culture*: Attempting to take over non-scientific disciplines and replace them with science.
7. *Devaluing the different*: Denigrating the importance, or even denying the legitimacy, of non-scientific disciplines.

According to Haack, every one of these indicators denotes some misunderstanding of science.

As the topic of scientism is impossible to treat in detail here, what I am concerned with in this section is only one component of scientism – whether there are other ways of knowing besides science. Philosopher Alex Rosenberg, while definitively denying the existence of God based on what we know about physics and biology, noted that atheists must indeed adopt scientism:

> This is the conviction that the methods of science are the only reliable ways to secure knowledge of anything; that science's description of the world is correct in its fundamentals; and that when "complete," what science tells us will not be surprisingly different from what it tells us today ... Science provides all the significant truths about reality, and knowing such truths is what real understanding is all about.

I will approach this topic through an example: Is the question about the existence of God a scientific one? Not surprisingly, there are those who think that it is. Richard Dawkins wrote that:

> God's existence or non-existence is a scientific fact about the universe, discoverable in principle if not in practice. If he existed and chose to reveal it, God himself could clinch the argument, noisily and unequivocally, in his favour. And even if God's existence is never proved or disproved with certainty one way or the other, available evidence and reasoning may yield an estimate of probability far from 50 per cent.

Dawkins then examined various arguments for God's existence, including the so-called ontological argument. This was first proposed in the eleventh century by St. Anselm of Canterbury in his *Proslogion*. There have been several attempts to reformulate this argument, which seems to go like this:

1. God exists in the understanding but not in reality (assumption).
2. Existence in reality is greater than existence in the understanding alone (premise).
3. A being having all of God's properties plus existence in reality can be conceived (premise).
4. A being having all of God's properties plus existence in reality is greater than God (from (1) and (2)).
5. A being greater than God can be conceived (from (3) and (4)).
6. It is false that a being greater than God can be conceived (from definition of "God").
7. Hence, it is false that God exists in the understanding but not in reality (from (1), (5), and (6)).
8. God exists in the understanding (premise, to which even the fool agrees).
9. Hence God exists in reality (from (7) and (8)).

Dawkins rejected this kind of argument by noting that:

> My own feeling, to the contrary, would have been an automatic, deep suspicion of any line of reasoning that reached such a significant conclusion without feeding in a single piece of data from the real world. Perhaps that indicates no more than that I am a scientist rather than a philosopher.

Philosopher Daniel Dennett provided a similar argument:

> you can't prove the existence of *anything* (other than an abstraction) by sheer logic. You can prove that *there is* a prime number larger than a trillion, and that *there is* a point at which the lines bisecting the three angles of any triangle all meet . . . but you can't prove that something that has effects in the physical world exists except by methods that are at least partly empirical.

And so did evolutionary biologist Jerry Coyne:

> With the notion of theistic god and a vernacular notion of "proof" in hand, we can disprove a god's existence in this way: *If a thing is claimed*

> *to exist, and its existence has consequences, then the absence of those consequences is evidence against the existence of the thing.* In other words, the absence of evidence – *if evidence should be there* – is indeed evidence of absence ... Many gods claimed to exist *should* have observable effects on the world. ... But the evidence isn't there: we see no miracles or miracle cures in today's world, much less any wondrous signs of a God who presumably wants us to know him.

In short, the argument is that if God existed, we should be able to test his existence by empirical means. Is this science or scientism? Are the claims of Dawkins, Dennett, and Coyne that the existence of God is a scientific question legitimate? Or do they try to apply the methods of science beyond its scope?

The first issue is how one defines God. There are various concepts of God that one might argue we can put under scientific test. For instance, let us assume that we consider God a benevolent and all-powerful entity. One might argue that there is no evidence that God exists because one has not observed the action of such an entity, such as those described in the scriptures. The design perceived in nature is evidence for the existence of God for some people, but others prefer to attribute it to natural selection. One might indeed claim that even if the design perceived in nature counted as evidence for the existence of God, it is only indirect evidence. People would be convinced if there was direct evidence, such as of the kind that Jesus Christ is said to have provided through miracles. The usual response of religious people to this is that absence of evidence is not evidence of absence, or simply that we cannot prove the negative. In other words, one might argue that the fact that we have not observed the actions of a benevolent and all-powerful God does not entail that God does not exist. One cannot establish the non-existence of something simply because there is no evidence of its existence. Such evidence may actually exist, but may have simply escaped our attention. This is actually a solid argument: That we have not found evidence for the existence of a transitional form in evolution does not entail that such a form has not existed. In the same sense, that we have not found direct evidence for the existence of God does not entail that God does not exist.

So far, so good. But there is a counter-argument here. One might argue, as Hume did (Chapter 2), that there is indeed evidence for the absence of a benevolent and all-powerful God. There is so much evil and cruelty in nature, and many people die every day because of it. Evil might thus stand as evidence for the absence of a benevolent and all-powerful God. If God is benevolent, why does he let evil exist? If he is all-powerful, why doesn't he stop it? If he is both benevolent and all-powerful, why does evil exist at all? Therefore, one might argue that it is not only that there is absence of evidence for the existence of God; there might also be evidence for his absence. That evil exists in the world is perceived by many as evidence for the non-existence of a benevolent and all-powerful god. As Darwin succinctly put it in a letter to Joseph Dalton Hooker: "What a book a Devil's chaplain might write on the clumsy, wasteful, blundering low & horridly cruel works of nature!" Of course, even if we accepted this argument, this would be based on indirect evidence as well. Whether design can be considered as evidence for the existence of God and evil as evidence for God's non-existence, they are both kinds of indirect evidence. Design and evil may be due to God's action or inaction, but they are not God himself.

Assuming that the existence of design and the existence of evil can be considered as empirical evidence for the existence and non-existence of God, respectively, is this scientific evidence? In my view, the arguments about the existence or non-existence of God are philosophical arguments, not scientific ones; and the evidence related to design and evil is not scientific evidence. The important point here is that empirical evidence is not necessarily scientific evidence. The latter has specific features that depend on the respective disciplines and that can be measured with whatever means and instruments these disciplines use. How can one measure design and evil? People might not even agree about how to define design and evil, let alone about how to measure them. This is the relatively subjective, but still rational and important, realm of philosophy. We are not in the realm of science, because design and evil are not the objects of scientific endeavor. Philosophical endeavor is of course important, but we should be conscious not to confuse it with science. Philosophical arguments are similar but also different from scientific arguments, in part because of the different kinds of empirical evidence on which they rely. We may have empirical evidence about the supernatural, such as the existence of design or evil, but this is not scientific evidence.

Science can refute claims of supernatural nature only if they have observable and scientifically measurable consequences. For instance, a study aimed to explore the impact of intercession (praying to a deity on behalf of others) on recovery from heart disease. In particular, it explored whether actually receiving intercessory prayer or being certain of receiving intercessory prayer was associated with uncomplicated recovery after surgery for coronary heart disease. Intercessory prayer was provided for 14 days, starting the night before surgery. Patients were randomly assigned to one of three groups: 604 of them received intercessory prayer after being informed that they may or may not receive prayer; 597 of them did not receive intercessory prayer also after being informed that they may or may not receive prayer; and 601 of them received intercessory prayer after being informed that they would receive prayer. The researchers found that major adverse events and mortality rates within 30 days from surgery were similar across all three groups. In the two groups who were uncertain about whether or not they would receive intercessory prayer, complications occurred in 52 percent of those who actually received prayer versus 51 percent of those who did not. In addition, complications occurred in 59 percent of patients who were certain of receiving prayer versus 52 percent of those who were uncertain of receiving intercessory prayer – those not receiving prayer actually did better than those who received prayer! Here are claims of supernatural nature that can have empirically testable consequences. But these consequences, that is, whether people died or had complications after their surgeries, are empirical evidence of a scientific nature – something that scientists can measure and study.

Realizing what scientists know, do not know, and cannot know depends on understanding the nature of the questions asked by scientists and the kinds of answers it is possible to provide. In general, science aims to answer questions about nature; the view that this is possible can be described as naturalism. The question then becomes how one approaches nature, and it is important to clearly distinguish between two types of naturalism: *metaphysical* naturalism and *methodological* naturalism. Metaphysical naturalism, also called philosophical or ontological naturalism, suggests that only natural entities exist and thus denies the existence of anything supernatural. Methodological naturalism does not deny the existence of supernatural entities, but nevertheless recognizes that one cannot study them and consequently that there is no reason to be concerned about them. Science is a practice of methodological naturalism: Whether a realm of the supernatural exists or not, it cannot be studied by the rational tools of science. Science does not deny the supernatural, but accepts that it does not have much to say about it, except for cases where there are observable consequences that scientists can study, such as in the case of intercession. Science is a method of studying nature, hence methodological naturalism. Science is also concerned with the metaphysics of nature (causes of natural phenomena), but does not always have something to say about them. Doing science is different from relying on it to develop a philosophical view. It is one thing to believe that God does not exist based on one's understanding of science, and another to say that science can definitively show that God does not exist. This entails more broadly that there are questions that cannot be answered by science.

Questions Not Answered by Evolutionary Theory

Scientific knowledge has specific characteristics; perhaps its most important one is that it has limits that correspond to the limits of human cognitive abilities. As a consequence, there will always be questions that scientific theories will not be able to answer, perhaps not even to address. Therefore, we must accept the limits of our cognitive abilities. Science is one way of knowing; there are other kinds of knowledge, such as religious and moral knowledge. Perhaps there is more. In any case, science cannot bring an end to our questions, worries, and concerns.

This book is about evolution, so I will explain which questions, I think, cannot be answered by evolutionary theory. This is particularly important because in my view public resistance to evolution is due to two main reasons. One has to do with the misuse of the theory in trying to answer questions that it cannot actually answer, already addressed in Chapter 2, and the other with deep human intuitions already addressed in Chapter 3. I think science and philosophy can help humans live a meaningful life. But at the same time, I also think we should be able to distinguish between the less subjective approach of science and the more subjective approach of philosophy. This is not to deny that scientific endeavor exhibits some elements of subjectivity; quite the contrary. However, the subjective elements in philosophy are more predominant and less easy to clarify.

In Chapter 2 I argued that scientists may have very different religious views and I described the views of Richard Dawkins, an atheist, Stephen Jay Gould, an agnostic, and Simon Conway Morris, a believer. I argued that their religious views notwithstanding, all three of them relied on science to answer questions that fall outside its realm. This is not wrong, as science can inform answers to such questions. However, I am concerned that evolutionary theory is unfairly blamed when it is being misused to answer questions it cannot actually answer, and it is actually blamed for too much. In order to explain why this is the case, I describe how I see the relationship between science and religion in analogy with the relationship between science and morality. In particular, I explain what implications biological science has for morality and I argue that the implications that evolutionary science has for religion are not more significant than that. Again, it is Christian religion I am referring to because evolutionary theory was developed within cultures in which Christianity was considered the dominant religion.

A main reason for antievolutionism is that evolutionary theory is often identified with a form of materialism, which is perceived as amoral or immoral. A common criticism is that evolutionary theory deprives human life of moral values and principles as it considers humans as just another animal species among all the others. In this view, if humans accept that they are just animals, as evolutionary theory suggests, they may start behaving like them: compete, kill, mate promiscuously, etc. What makes things even worse is the outbreak of what has been called *militant modern atheism*, which provides a critique of religious fundamentalism, but also often stands as an attack against all religion. When the criticism of religious fundamentalism is extended to all religious attitudes, and when acceptance of evolution is promoted as the only rational alternative to the big questions, a conflict arises. A consequence is that evolutionary theory is identified with atheism or materialism and thus many people become afraid of it and its consequences before they even have the chance to understand what it is about.

Generally speaking, science has implications both for morality and for religion. However, although one could be informed by science to make relevant decisions, one cannot base these decisions solely on science. Morality and religion also have to do with worldviews and philosophical perspectives. These can be enriched by science in various ways, but science cannot guide them because decisions about what is bad or wrong, as well as about whether human life has an inherent purpose or not, are made on a subjective basis. For example, we might decide that either humans only or all organisms should be included in the moral community. There are several arguments for and against each choice. Knowledge of which beings are sentient or not might be useful to consider, but it could not alone point to any decision about this. Other arguments, provided by the various ethical theories, are also important to consider. Eventually, there will always be a counter-argument about what we decide, and whether we will take it into account or ignore it is rather subjective.

Morality is not threatened by science in any way, because science can only inform, but not guide, moral choices and decisions. Scientific knowledge may make people question moral values and norms, but this does not mean that the latter will persist only if the former is rejected. For example, clinical trials could be considered immoral if healthy human subjects were given experimental drugs without their consent. However, this does not mean that we should reject clinical trials altogether as a practice; rather, we should be careful in their implementation and ensure that they are used appropriately and with respect for participants.

Ethical decisions should be made after having obtained a good understanding of the respective scientific knowledge, but the final decision about what is moral or not requires more than scientific knowledge alone. In other words, science can make important contributions to decisions about ethics, but it cannot guide them. Some people might acquire the contemporary scientific knowledge – e.g., that double-blind clinical trials might provide us with important conclusions about drugs and their effects – but nevertheless decide not to perform them because of, e.g., their particular cultural characteristics. It is their right to do so, and they should not blame other cultures for adopting them, as long as their implementation respects human life. Of course, such decisions at the community level can only be made if a consensus is reached.

Religion is perhaps crucial not only at a community level, but also at a more personal level. Nevertheless, people can make decisions in a similar manner. People should be aware of contemporary scientific knowledge and then decide if it has implications for their religion and how serious these implications are. Some people might decide to change their religious views, others might not. No matter what happens to one's religious beliefs, science can inform but not guide such a decision. One might decide to sustain one's religious beliefs, even despite an apparent incompatibility with science. What is important is that one is aware of this knowledge; then a conscious decision can be made. A good understanding of science does not entail a rejection of religion. This is a philosophical decision that might be affected by science, but not guided by it. The problem is when people see some incompatibility between science and religion, and as a result decide to reject science. It is a huge mistake to reject the understanding of the natural world that science has provided because it conflicts with our religious beliefs and worldviews. Science does not rationally entail a rejection of religion, but rejecting science because it is incompatible with religion is irrational.

One should think hard about these issues before reaching conclusions. Whatever these conclusions are, they are personal choices. Science *per se* is not religious or atheistic, moral or immoral. Scientists can be either of these. Therefore, it is scientists that should be criticized about how they act and what conclusions they make. And if we disagree with a religious person or with an atheist, we should neither reject nor blame science for this. Disagreement is a healthy endeavor. If I use science to justify my religion or to justify my atheism, you should not blame science for my views, but me. Similarly, if I use science to support or reject the use of stem cells, I am solely responsible for this. A fertilized ovum will not develop to an embryo unless it is implanted in a uterus. This is a fact of life. Now, whether I consider a fertilized egg as a human being because it has the potential to develop into an adult once it is implanted, or as just a bunch of cells with some potential that will never develop to an adult if it is not implanted is, in my view, a subjective decision. More generally, how one relies on science to make philosophical conclusions cannot have implications for science.

Whether scientists are atheists, like Dawkins, or religious believers, like Conway Morris, or agnostics, like Gould, or whatever else, evolutionary theory itself has nothing to do with it. None of these conclusions stems from scientific understanding; rather, it stems from a personal interpretation of scientific understanding within different contexts. What these people should be saying is not whether God exists or not, but rather that these are the inferences they make based on their own understanding of evolutionary theory. The fact that Dawkins, Conway Morris, and Gould made such different inferences shows that evolutionary theory itself has nothing to say about religion, even though it can of course inform our respective decision to believe in God or not. Whether you disagree with their inferences or not, it has no implications for evolutionary theory itself.

To better illustrate this point, let us turn to religion, which has been criticized as supporting violence and other bad actions. But is it really religion that is to blame or religious fundamentalists? When people are killing each other in the name of their religion, why is religion to blame, and not the people themselves? It might make sense to claim that if there were no cars, there would be no traffic accidents – and thousands die every year in traffic accidents. However, in most such cases the drivers, and not the cars, are to blame. The point I want to make is that all of us are responsible for what we say, write, or do. If I drink at a party, I can decide not to drive my family home. If I drink so much that I am not able to make such a decision, this is my fault – I cannot blame my favorite red wine for this. Similarly, if my scientific knowledge makes me believe that God does or does not exist, I am responsible for such a view. I have the right to defend it, but I must take responsibility for it; no one should blame evolutionary theory for the conclusions I reach. If you disagree with what I am writing here and with the conclusions I will soon make from evolutionary theory, you have the right to reject them. There is no point in rejecting evolutionary theory, because it is not responsible for my atheism, agnosticism, or religiosity. This is my decision; my choice; my responsibility.

To summarize: I believe that our decision to embrace or reject religion, or to respect or ignore moral values, can be informed by science; however, the respective arguments that support it are philosophical, not scientific. Therefore, there are important questions that science alone cannot answer. We can decide to draw on science in order to provide our own answers, but we should always remember to clearly distinguish between scientific knowledge and our religious or moral views.

Concluding Remarks

The Implications of Evolutionary Theory for Human Life

I believe that the problems with the public acceptance of evolution do not have only to do with understanding science, but also with respecting one another's worldviews. I try to respect those who disagree with me. We do not always do this, and I believe that misallodoxy (a Greek word: misos: hate; allo: other; doxa: belief), that is, hating other people because they hold different beliefs than one does, is a major problem in human societies and in part responsible for the so-called evolution wars. The problem in the case of evolution and religion is that we fail to respect the views of those who disagree with us. Militant atheists fail to respect the decision of religious people to believe; religious fundamentalists fail to respect the decision of irreligious people or atheists not to believe. Militant atheists blame religion and religious fundamentalists blame science. I think they are all wrong. Writing on the science side, I want to conclude that evolutionary theory influences but does not guide atheism, as well as that it has implications for religion but does not necessarily hurt it. Evolutionary theory provides a deep, coherent understanding of our natural world; it does not have much to say about the supernatural, although one can draw several philosophical conclusions about it. As all science, evolutionary theory is a human construct and indeed a successful one given how many questions it can answer and how many applications it has. How we use it to make claims about anything beyond the natural world is our own problem, and not one of the theory itself.

Personally, I have made the choice to worry about whatever my human brain can process. Science is knowledge of this kind. Whether God exists or not, I feel that I do not have much to say; I find myself somewhere between Richard Dawkins and Stephen Jay Gould in the continuum I described in

Chapter 2. But I think that people should believe whatever they want. They may be religious or irreligious; they may believe in God or not believe in God. But what matters, in my view, is having respect for the views of others. I should note at this point that my personal views notwithstanding, I certainly understand the need for religion. Michael Ruse and Philip Kitcher, two atheist philosophers, have written book-length treatises about how there is always a place for religion in the rational world of science. Instead of describing their views in detail, I prefer to quote their conclusions, with which I could not agree more:

> Can a Darwinian be a Christian? Absolutely! Is it always easy for a Darwinian to be a Christian? No, but whoever said that the worthwhile things in life are easy? Is the Darwinian obligated to be a Christian? No, but try to be understanding of those who are. Is the Christian obligated to be a Darwinian? No, but realize how much you are going to foreswear if you do not make the effort, and ask yourself seriously (if you reject all forms of evolutionism) whether you are using your God-given talents to the full. (Ruse)

> There is truth in Marx's dictum that religion ... is the opium of the people, but the consumption should be seen as medical rather than recreational ... Genuine medicine is needed, and the proper treatment consists of showing how lives can matter ... In addressing these issues we may discover that the deliverances of reason can be honored without ignoring the most important human needs – and going beyond supernaturalism, that we can live with Darwin, after all. (Kitcher)

Given these considerations, is there any purpose and meaning in our evolving world? I take evolutionary theory to suggest that there is no inherent purpose and meaning in our world. But this is my own view; if it is wrong, this is my fault – it has no consequences for evolutionary theory itself. Even if you agree with me that this is the case, I do not think this means that one cannot find purpose and meaning in life in this evolving world. In contrast, I see our ability to consider such questions as an evolutionary outcome – some might say a triumph of evolution, given that our closest relatives do not seem to have such concerns. Evolutionary theory does not deprive life of meaning; in contrast, it shows us that we are fortunate to be able to feel happiness, satisfaction, to set goals, and to try to fulfill them. Most animals just kill other organisms to feed on their tissues, reproduce, and die. We do that too, but we are also in a position to be aware of that, as well as that there is more we can do: think, communicate, and philosophize. Biologically speaking, we are primates; but perhaps in contrast with most primates and other animals, we have a wider scope of experiences that stem from our biological hypostasis, but that also go beyond that. We have culture, morals, and much more. In many cases, but unfortunately not for all people on Earth, we do not just survive; we live.

I also think that evolutionary theory has another important implication for human life – again, this is my view, and if it is wrong there are no consequences for the theory itself. I think that those who understand evolutionary theory realize that we humans should not be arrogant. I take evolutionary theory to suggest that we have no special place in this world and thus no justification to believe that we can rule it in any way we like. We are not a special or select species, but just one species among so many; we belong to a short and recent node in the evolutionary network of life. We have no right to change the world any way we like, just because (we think) we can. Most interestingly, we do not have the power we think we have in order to do this. Although the human population at the global level increases every day, millions of people still die from infectious disease. We are not all-powerful. Yet we cause enormous change in the natural world. This is not necessarily a bad thing to do; photosynthetic organisms transformed the atmosphere of the Earth by producing molecular oxygen without which perhaps we would never have evolved. I do not know if the changes we cause will have a good or a bad outcome; history will tell, even though as I write these lines the concerns about anthropogenic climate change are many. But we have no reason and no rational justification to believe we can change the world any way we like just because this is our world. It is not! We are just a very minor component of this world. We should not be arrogant, but modest.

In accepting that there is no inherent purpose in the world and that we should not be so arrogant as to believe that we have any special place reserved in it, we can decide how to live our lives and what matters most. I strongly believe we can find meaning and purpose in life. I believe that life becomes meaningful through sentiments. Given the violence that is inherent in our world, I am happy to be able not to worry about my survival and to be able to write these lines. I find enormous meaning in my life in doing everyday things with my wife and our children. I feel extremely rich of sentiments when they all express their love for me. I find meaning in writing this book and sharing my thoughts and understanding of the world with people all over the globe. These different kinds of meaning contribute to the purpose I find in life: live long, love people, enjoy life, and feel full of sentiments and happiness when I get older and the final countdown begins. Evolutionary theory has nothing to say about all this, but these are implications I have drawn from my own understanding of evolution. So, if I may make a suggestion, it is this: First try to understand evolutionary theory; then draw your own implications for your life, and find your own meaning and purpose.

Summary of Common Misunderstandings

Common misunderstandings about evolution and responses stemming from this book:

Organisms like humans and bacteria are too different to be related. Despite the morphological diversity of organisms, they are all related and belong to the same evolutionary network of life. Their crucial similarities at the morphological, molecular, and developmental levels (genetic code, DNA, cells, metabolism, reproduction, etc.) stand as strong evidence for their common ancestry.

Some characteristics exist in quite similar forms in very different species and thus reflect the same basic design. Despite the morphological diversity among organisms, there are characteristics that exist in many of them that have appeared again and again in the course of evolution. These similarities may have evolved independently through selection for characteristics that are beneficial under similar environmental conditions. However, it is also possible that these similar characteristics are homologous at some deeper (molecular or developmental) level.

Large morphological changes in evolution are impossible because they require numerous changes in DNA sequences. The large morphological diversity among organisms may make it difficult for people to realize how species with very different forms might be related. However, large morphological transitions in evolution are possible through minor changes at the molecular and the developmental levels. This happens because much as organisms differ morphologically, they also have similarities in their developmental essences due to their common ancestry. Very different morphological characteristics can arise when even minor changes in their common developmental processes occur.

The marvelous adaptations that organisms exhibit can only be the outcome of design. Organisms sometimes exhibit marvelous adaptations that seem to be the outcome of intentional design. However, this is an illusion because they also exhibit features that are not at all advantageous. The characteristics that perform functions can be the outcome of natural selection. These characteristics confer, or have conferred in the past, an advantage to the survival and reproduction of their bearers, and it is because of this advantage that they exist.

Organisms will acquire the characteristics necessary for their survival if there is sufficient time for this to occur. Which characteristics the various organisms will come to possess does not depend on their needs, but on the changes that will take place in their DNA and the subsequently emerging phenotypic variation. Whether some phenotypes will have an advantage over others or not is a different issue. Some necessary features might never emerge and the individuals that "need" them might simply die out.

Evolution is a random process. Evolution is not a random process. However, there are indeed evolutionary processes in which chance plays a major role, called stochastic processes. A main feature of evolution is contingency: Evolutionary outcomes are contingent *per se*, that is, they are unpredictable and one cannot know in advance which one among various possible outcomes will come to be. Once an outcome occurs, all future outcomes are contingent upon it, because it has thereafter determined the path along which evolution will proceed. Contingency is incompatible with any notion of intentional design.

There has not been sufficient time for evolution to occur. The time that has elapsed since the emergence of our planet has been enormous and more than enough for the evolution of life to occur. Much of this process has resulted in the extinction of most species that have ever appeared on Earth.

We cannot establish that evolution has occurred because we cannot observe past events. The ample indirect evidence for evolution is more than sufficient to establish the fact of evolution through natural processes. There is no need for us to have observed evolutionary processes in all their detail, because the traces that these processes have left are sufficient for understanding. Of course, there are details we still do not know, and that we may never know. But these do not affect our understanding of evolution.

Evolutionary theory leads to atheism. Evolutionary theory shows that we are part of the organic world, and that our species has evolved relatively recently. Whether this leads one to atheism, agnosticism, or theism, or any other worldview, is a matter of personal interpretation and inferences, not something that evolutionary theory itself entails.

Evolutionary theory deprives life of meaning. Evolutionary theory supports the conclusion that we are not special, but only one species among numerous others. Whether this renders human life meaningless or a triumph of evolution is a matter of personal interpretation.

References

Preface

On the evidence for evolution: Coyne, J. A. (2009). *Why Evolution is True*. Oxford: Oxford University Press; Dawkins, R. (2009). *The Greatest Show on Earth: The Evidence for Evolution*. London: Bantam Press; Prothero, D. R. (2017). *Evolution: What the Fossils Say and Why it Matters* (2nd edition). New York: Columbia University Press.

"Nothing in biology makes sense . . .": Dobzhansky, T. (1973). Nothing in biology makes sense except in the light of evolution. *The American Biology Teacher* 35(3): 125–129.

Creationism and intelligent design: Numbers, R. L. (2006). *The Creationists: From Scientific Creationism to Intelligent Design*. Cambridge MA: Harvard University Press; Pigliucci, M. (2002). *Denying Evolution: Creationism, Scientism, and the Nature of Science*. Sunderland, MA: Sinauer Associates.

Blind watchmaker and tinkerer metaphors: Dawkins, R. (2006 [1986]). *The Blind Watchmaker*. London: Penguin Books; Jacob, F. (1977). Evolution and tinkering. *Science* 196(4295): 1161–1166.

Chapter 1

"Make respondents feel uncomfortable": Rughinis, C. (2011). A lucky answer to a fair question: Conceptual, methodological, and moral implications of including items on human evolution in scientific literacy surveys. *Science Communication* 33: 501–532.

"Forced to choose": Elsdon-Baker, F. (2015). Creating creationists: The influence of "issues framing" on our understanding of public perceptions of clash narratives between evolutionary science and belief. *Public Understanding of Science* 24: 422–439.

Articles on the acceptance of evolution in *Science*: Miller, J. D., Scott, E. C., and Okamoto, S. (2006). Public acceptance of evolution. *Science* 313(5788): 765–766; Hameed, S. (2008). Bracing for Islamic creationism. *Science* 322: 1637.

On the distinction between *believe in* and *believe about*: McCain, K. and Kampourakis, K. (2018). Which question do polls about evolution and belief really ask, and why does it matter? *Public Understanding of Science* 27(1): 2–10.

Chapter 2

On evolution and design: Ruse, M. (2004). *Darwin and Design: Does Evolution Have a Purpose?* Cambridge MA: Harvard University Press.

Paley and Hume: Paley, W. (2006 [1802]). *Natural Theology or Evidence of the Existence and Attributes of the Deity, Collected from the Appearances of Nature*. Oxford: Oxford University Press; Hume, D. (1993 [1779/1777]). *Dialogues Concerning Natural Religion and Natural History of Religion*. Oxford: Oxford University Press.

The studies discussed are Evans, M. E. (2001). Cognitive and contextual factors in the emergence of diverse belief systems: Creation versus evolution. *Cognitive Psychology* 42: 217–266; Kelemen, D. (2003). British and American children's preferences for teleo-functional explanations of the natural world. *Cognition* 88: 201–221; Kelemen, D. (2004). Are children "intuitive theists"? Reasoning about purpose and design in nature. *Psychological Science* 15(5): 295–301; Kelemen, D. and DiYanni, C. (2005). Intuitions about origins: Purpose and intelligent design in children's reasoning about nature. *Journal of Cognition and Development* 6(1): 3–31.

On the blind watchmaker: Dawkins, R. (2006 [1986]). *The Blind Watchmaker*. London: Penguin Books.

Studies on what scientists think about religion: Ecklund, E. H. (2010). *Science vs. Religion: What Scientists Really Think*. Oxford: Oxford University Press; Ecklund, E. H., Johnson, D. R., Vaidyanathan, B., Matthews, K. R. W., Lewis S. W., Thomson, Jr. R. A., and Di, D. (2019). *Secularity and Science: What Scientists Around the World Really Think About Religion*. Oxford: Oxford University Press.

On the views of Dawkins, Gould, and Conway Morris: Dawkins, R. (2006). *The God Delusion*. London: Bantam Press; Conway Morris, S. (2003). *Life's Solution: Inevitable Humans in a Lonely Universe*. Cambridge: Cambridge University Press; Gould, S. J. (1999). *Rocks of Ages: Science and Religion in the Fullness of Life*. New York: Ballantine Books.

On knowledge and belief: Audi, R. (2011). *Epistemology: A Contemporary Introduction to the Theory of Knowledge* (3rd ed.). New York: Routledge.

Chapter 3

On concepts and conceptual change: Arabatzis, T. (2019). What are scientific concepts? In K. McCain and K. Kampourakis (Eds.) *What Is Scientific Knowledge? An Introduction to Contemporary Epistemology of Science*. New York: Routledge, pp. 85–99; Vosniadou, S. (Ed.) (2013). *International Handbook of Research on Conceptual Change* (2nd ed.). New York: Routledge.

"Interesting study as an example": Goldberg, R. F. and Thompson-Schill, S. L. (2009). Developmental "roots" in mature biological knowledge. *Psychological Science* 20(4): 480–487.

The studies on teleology discussed are reported in: Kelemen, D. (1999). The scope of teleological thinking in preschool children. *Cognition* 70: 241–272; Keil, F. C. (1992). The origins of an autonomous biology. In M. R. Gunnar and M. Maratsos (Eds.) *Modularity and Constraints in Language and Cognition: Minnesota Symposium on Child Psychology*, Vol. 25. Hillsdale, NJ: Erlbaum, pp. 103–138; Kelemen, D. (1999). Why are rocks pointy? Children's preference for teleological explanations of the natural world. *Developmental Psychology* 35: 1440–1452; Greif, M., Kemler-Nelson, D., Keil, F. C., and Guiterrez, F. (2006). What do children want to know about animals and artifacts? Domain-specific requests for information. *Psychological Science* 17(6): 455–459.

The studies on essentialism discussed are reported in: Gelman, S. A. (2003). *The Essential Child: Origins of Essentialism in Everyday Thought*. Oxford: Oxford University Press; Asher, Y. M. and Kemler-Nelson, D. G. (2008). Was it designed to do that? Children's focus on intended function in their conceptualization of artifacts. *Cognition* 106: 474–483; Keil, F. C. (1989). *Concepts, Kinds and Cognitive Development*. Cambridge, MA: MIT Press.

On the essences of artifacts and organisms: Bloom, P. (2004). *Descartes' Baby: How the Science of Child Development Explains What Makes Us Human*. New York: Basic Books; Walsh, D. (2006). Evolutionary essentialism. *British Journal for the Philosophy of Science* 57(2): 425–448.

On teleology and essentialism as obstacles to understanding evolution: Kelemen, D. (2012). Teleological minds: How natural intuitions about agency and purpose influence learning about evolution. In K. Rosengren, S. Brem, E. M. Evans, and G. M. Sinatra (Eds.) *Evolution Challenges: Integrating Research and Practice in Teaching and Learning about Evolution*. Oxford: Oxford University Press, pp. 66–92; Gelman, S. A. and Rhodes, M. (2012). "Two-thousand years of stasis": How psychological essentialism impedes evolutionary understanding. In K. Rosengren, S. Brem, E. M. Evans, and G. M. Sinatra (Eds.) *Evolution Challenges: Integrating Research and Practice in Teaching and Learning about Evolution*. Oxford: Oxford University Press, pp. 3–21.

On the account of teleology followed, see: Lennox, J. G. and Kampourakis, K. (2013). Biological teleology: The need for history. In K. Kampourakis (Ed.) *The Philosophy of Biology: A Companion for Educators*. Dordrecht: Springer, pp. 421–454; Kampourakis, K. (2020). Students' "teleological misconceptions" in evolution education: Why the underlying design stance, not teleology per se, is the problem. *Evolution: Education and Outreach*, 13(1). DOI: 10.1186/s12052-019-0116-z.

Chapter 4

The account of the development of Darwin's theory is largely, but not exclusively, based on the following books.

Bowler, P. J. (2013). *Darwin Deleted: Imagining a World Without Darwin*. Chicago, IL: University of Chicago Press.

Endersby, J. (2009). *Charles Darwin: On the Origin of Species*. Cambridge: Cambridge University Press.

Hodge, J. and Radick, G. (Eds.) (2009). *The Cambridge Companion to Darwin* (2nd ed.). Cambridge: Cambridge University Press.

Kohn, D. (Ed.) (1985). *The Darwinian Heritage*. Princeton, NJ: Princeton University Press.

Numbers, R. N. & Kampourakis, K. (Eds.) (2015). *Newton's Apple and Other Myths about Science*. Cambridge, MA: Harvard University Press.

Ospovat, D. (1981). *The Development of Darwin's Theory: Natural History, Natural Theology and Natural Selection, 1838–1859*. Cambridge: Cambridge University Press.

Ruse, M. and Richards, R.J. (Eds.) (2009). *The Cambridge Companion to the "Origin of Species."* Cambridge: Cambridge University Press.

Secord, J.A. (2000). *Victorian Sensation: The Extraordinary Publication, Reception, and Secret Authorship of Vestiges on Natural History of Creation*. Chicago, IL: University of Chicago Press.

Spencer, N. (2009). *Darwin and God*. London: SPCK.

On the Huxley–Wilberforce debate: Lucas, J.R. (1979). Wilberforce and Huxley: A legendary encounter. *The Historical Journal* 22(2): 313–330; Brooke, J.H. (2001). The Wilberforce–Huxley debate: Why did it happen? *Science & Christian Belief* 13(2): 127–141.

Darwin's writings, as well as the reviews of the *Origin* discussed, are available at: http://darwin-online.org.uk. Darwin's correspondence is available at www.darwinproject.ac.uk.

Chapter 5

On the human genome and design: Avise, J.C. (2010). *Inside the Human Genome: A Case for Non-Intelligent Design*. Oxford: Oxford University Press, pp. 108–112.

On microbial evolution and the universal common ancestor: Margulis, L. (1998). *Symbiotic Planet: A New Look at Evolution*. London: Basic Books; Margulis, L. and Sagan, D. (2002). *Acquiring Genomes: A Theory of the Origins of Species*. New York: Basic Books; Koonin, E.V. (2011). *The Logic of Chance: The Nature and Origin of Biological Evolution*. Upper Saddle River, NJ: FT Press; Doolittle, W.F. and Brunet, T.D. (2016). What is the tree of life? *PLoS Genetics* 12(4): e1005912; Koonin, E.V. and Novozhilov, A.S. (2017). Origin and evolution of the universal genetic code. *Annual Review of Genetics* 51: 45–62.

For a comprehensive discussion of how evolutionary trees are constructed and read, see: Baum, D. and Smith, S. (2013). *Tree Thinking: An Introduction to Phylogenetic Biology*. Greenwood Village, CO: Roberts and Company Publishers.

On digits and insect appendages: Hall, B. K. (2003). Descent with modification: The unity underlying homology and homoplasy as seen through an analysis of development and evolution. *Biological Reviews of the Cambridge Philosophical Society* 78: 409–433; Wagner, G. (2007). The developmental genetics of homology. *Nature Reviews Genetics* 8: 473–479.

On homologies and homoplasies: Carroll, S. B. (2005). *Endless Forms Most Beautiful: The New Science of Evo-Devo*. New York: W.W. Norton, pp. 61–72; McGhee, G. R. (2011). *Convergent Evolution: Limited Forms Most Beautiful*. Cambridge, MA: MIT Press.

On deep homology: Shubin, N., Tabin, C., and Carroll, S. (2009). Deep homology and the origins of evolutionary novelty. *Nature* 457: 818–823.

On the origin of multicellularity: Knoll, A. H. (2011). The multiple origins of complex multicellularity. *Annual Review of Earth and Planetary Sciences* 39: 217–239; Libby, E., Conlin, P. L., Kerr, B., & Ratcliff, W. C. (2016). Stabilizing multicellularity through ratcheting. *Philosophical Transactions of the Royal Society B: Biological Sciences* 371(1701): 20150444.

On whales and bats: Thewissen, J. G., Cohn, M. J., Stevens, L. S., Bajpai, S., Heyning, J., and Horton, W. E., Jr. (2006). Developmental basis for hind-limb loss in dolphins and origin of the cetacean body plan. *Proceedings of the National Academy of Sciences USA* 103(22): 8414–8418; Sears, K. E., Behringer, R. R., Rasweiler, J. J., and Niswander, L. A. (2006). Development of bat flight: morphologic and molecular evolution of bat wing digits. *Proceedings of the National Academy of Sciences USA* 103(17): 6581–6586.

On changes in development that affect evolution: Arthur, W. (2011). *Evolution: A Developmental Approach*. Oxford: Wiley-Blackwell.

On the inversion of the dorso-ventral axis of arthropods: De Robertis, E. M. (2008). Evo-devo: Variations on ancestral themes. *Cell* 132: 185–195.

Chapter 6

On adaptation, see: Sober, E. (1993 [1984]). *The Nature of Selection: Evolutionary Theory in Philosophical Focus*. Chicago, IL: University of Chicago Press; Brandon, R. N. (1990). *Adaptation and Environment*. Princeton, NJ: Princeton University Press; Reeve, H. K. and Sherman, P. W. (1993). Adaptation and the goals of evolutionary research. *Quarterly Review of Biology* 68(1): 1–32; Mayr, E. (2002). *What Evolution Is*. London: Weidenfeld and Nicolson; Gould, S. J.

and Vrba, E.S. (1982). Exaptation: A missing term in the science of form. *Paleobiology* 8(1): 4–15.

On selection for and selection of, and the toy example, see: Sober, E. (1993 [1984]). *The Nature of Selection: Evolutionary Theory in Philosophical Focus*. Chicago, IL: University of Chicago Press.

On selection *for* and *against*, see: Depew, D. (2013). Conceptual change and the rhetoric of evolutionary theory: "Force talk" as a case study and challenge for science pedagogy. In K. Kampourakis (Ed.) *The Philosophy of Biology: A Companion for Educators*. Dordrecht: Springer, pp. 121–144.

On stochastic processes: Millstein, R.L., Skipper, R.A., and Dietrich, M.R. (2009). (Mis)interpreting mathematical models: Drift as a physical process. *Philosophy & Theory in Biology* 1: e002, DOI:10.3998/ptb.6959004.0001.002; Skipper, R.A. (2006). Stochastic evolutionary dynamics: Drift versus draft. *Philosophy of Science* 73(5): 655–665; Nei, M. (2013). *Mutation-Driven Evolution*. Oxford: Oxford University Press.

Picket fence example: McShea, D.W. and Brandon, R.N. (2010). *Biology's First Law: The Tendency for Diversity and Complexity to Increase in Evolutionary Systems*. Chicago, IL: University of Chicago Press.

On evolutionary contingency: Gould, S.J. (2000 [1989]). *Wonderful Life: The Burgess Shale and the Nature of History*. London: Vintage; Kampourakis, K. (2018). *Turning Points: How Critical Events Have Driven Evolution, Life and Development.* Amherst, NY: Prometheus Books; Losos, J. (2017). *Improbable Destinies: Fate, Chance and the Future of Evolution.* New York: Riverhead Books.

On adaptive radiation: Losos, J.B. (2010). A tale of two radiations: Similarities and differences in the evolutionary diversification of Darwin's finches and Greater Antillean *Anolis* lizards. In P.R. Grant and B.R. Grant (Eds.) *In Search of the Causes of Evolution: From Field Observations to Mechanisms*. Princeton, NJ: Princeton University Press, pp. 309–331.

On evo-devo and speciation: Minelli, A. and Fusco, G. (2012). On the evolutionary developmental biology of speciation. *Evolutionary Biology* 39: 242–254.

On species selection: Jablonski, D. (2008). Species selection: Theory and data. *Annual Review of Ecology, Evolution and Systematics* 39: 501–524.

On the K–Pg extinction: Cleland, C.E. (2002). Methodological and epistemic differences between historical science and experimental science. *Philosophy of Science*

69: 474–496; Cleland, C. E. (2011). Prediction and explanation in historical natural science. *British Journal of Philosophy of Science* 62: 551–582; Cleland, C. (2020). Is it possible to scientifically reconstruct the history of life on Earth? The biological sciences and deep time. In K. Kampourakis and T. Uller (Eds.) *Philosophy of Science for Biologists*. Cambridge: Cambridge University Press.

Chapter 7

On the evidence that people can be religious and accept science: Ecklund, E. H. and Scheitle, C. P. (2017). *Religion vs. Science: What Religious People Really Think*. Oxford: Oxford University Press.

On the arguments of antievolutionists: Miller, K. R. (2009). *Only a Theory: Evolution and the Battle for America's Soul*. New York: Penguin Books; Pennock, R. T. (1999). *The Tower of Babel: The Evidence Against the New Creationism*. Cambridge, MA: MIT Press.

On nature of science: Kampourakis, K. and McCain, K. (2020). *Uncertainty: How it Makes Science Advance*. New York: Oxford University Press; Firestein, S. (2015). *Failure: Why Science is So Successful*. New York: Oxford University Press; Firestein, S. (2012). *Ignorance: How It Drives Science*. New York: Oxford University Press.

On scientific understanding: De Regt, H. (2017). *Understanding Scientific Understanding*. New York: Oxford University Press.

Virtues of a good theory: McMullin, E. (2008). The virtues of a good theory. In S. Psillos and M. Curd (Eds.) *The Routledge Companion to Philosophy of Science*. New York: Routledge, pp. 498–508.

On Tiktaalik: Shubin, N. (2008). *Your Inner Fish: The Amazing Discovery of our 375-Million-Year-Old Ancestor*. London: Penguin Books.

On scientism: Haack, S. (2017). *Scientism and Its Discontents*. n.p.: Rounded Globe; Boudry, M. and Pigliucci, M. (Eds.) (2018). *Science Unlimited? The Challenges of Scientism*. Chicago, IL: University of Chicago Press; Rosenberg, A. (2011). *The Atheist's Guide to Reality: Enjoying Life without Illusions*. New York: W.W. Norton; Dawkins, R. (2006). *The God Delusion*. London: Bantam Press; Dennett, D. (2006). *Breaking the Spell: Religion as a Natural Phenomenon*. New York: Viking; Coyne, J. (2015). *Faith vs. Fact: Why Science and Religion are Incompatible*. New York: Viking, pp. 202–204.

On the ontological argument: Plantinga, A. (1967). *God and Other Minds*. Ithaca, NY: Cornell University Press, quoted in Oppy, G., Ontological arguments. *The Stanford Encyclopedia of Philosophy* (spring 2019 edition), https://plato.stanford.edu/archives/spr2019/entries/ontological-arguments.

Effect of prayer on recovery: Benson, H., Dusek, J. A., Sherwood, J. B., et al. (2006). Study of the Therapeutic Effects of Intercessory Prayer (STEP) in cardiac bypass patients: A multicenter randomized trial of uncertainty and certainty of receiving intercessory prayer. *American Heart Journal* 151(4): 934–942.

On metaphysical and methodological naturalism: Giere, R. N. (2006). *Scientific Perspectivism*. Chicago, IL: University of Chicago Press.

Concluding Remarks

Quotations are from: Ruse, M. (2001). *Can a Darwinian be a Christian? The Relationship between Science and Religion*. Cambridge: Cambridge University Press, p. 217; Kitcher, P. (2007). *Living with Darwin*. New York: Oxford University Press, pp. 165–166.

Figure Credits

1.1 Image © Williammpark.
1.2 Data from the Eurobarometer 224: Europeans, Science and Technology report.
1.3 Data from the Eurobarometer 225: Social Values, Science and Technology report.
1.4 Data from the Eurobarometer 224: Europeans, Science and Technology report and the Eurobarometer 225: Social Values, Science and Technology report.
1.5 Data from www.gallup.com/poll/21814/evolution-creationism-intelligent-design.aspx.
1.6 Data from the following polls: Gallup May 8–11, 2014; May 3–6, 2012; December 10–12, 2010; Pew April 28 to May 12, 2009; March 21 to April 8, 2013; August 15–25, 2014.
1.7 Data from the Ipsos 2011 study.
1.8 Data from the Ipsos 2011 study.
1.9 Data from the Ipsos 2011 study.
3.1 Image © Simon Tegg.
3.2 Image © Simon Tegg.
4.1 Photograph by Maull & Fox, c. 1857. (DAR 225:175). With permission from the syndics of Cambridge University Library.
5.1 Image adapted, © Alashi.
5.2 Reproduced from Long, J. (2012). Evolution, missing links and climate change: recent advances in understanding transformational macroevolution. In A. Poiani (Ed.) Pragmatic Evolution: Applications of Evolutionary Theory. Cambridge: Cambridge University Press, 23–36, with permission from Cambridge University Press.

5.3 Image adapted from "300 vector silhouettes of animals (mammals, birds, fish, insects)" © Shutterstock.com/ntnt; "Big collection of vector" © Shutterstock.com/Pavel K; "66 pieces of detailed vectoral (fish silhouettes) © Shutterstock.com/ pinare; "vector animals silhouettes" © Shutterstock.com/lilac.

5.4 Image adapted from "300 vector silhouettes of animals (mammals, birds, fish, insects)" © Shutterstock.com/ntnt; "Big collection of vector" © Shutterstock.com/Pavel K; "66 pieces of detailed vectoral (fish silhouettes) © Shutterstock.com/pinare; "vector animals silhouettes" © Shutterstock.com/lilac.

5.5 Image adapted from "300 vector silhouettes of animals (mammals, birds, fish, insects)" © Shutterstock.com/ntnt; "Big collection of vector" © Shutterstock.com/Pavel K; "66 pieces of detailed vectoral (fish silhouettes) © Shutterstock.com/pinare; "vector animals silhouettes" © Shutterstock.com/lilac.

5.7 Image adapted from "300 vector silhouettes of animals (mammals, birds, fish, insects)" © Shutterstock.com/ntnt; "Big collection of vector" © Shutterstock.com/Pavel K; "66 pieces of detailed vectoral (fish silhouettes) © Shutterstock.com/pinare; "vector animals silhouettes" © Shutterstock.com/lilac.

5.8 Adapted from Coyne, J. A. (2009). *Why Evolution is True*. Oxford: Oxford University Press, p. 54. Image © Simon Tegg.

5.9 Adapted from "300 vector silhouettes of animals (mammals, birds, fish, insects)" © Shutterstock.com/ntnt; "Big collection of vector" © Shutterstock.com/Pavel K; "66 pieces of detailed vectoral (fish silhouettes) © Shutterstock.com/pinare; "vector animals silhouettes" © Shutterstock.com/lilac.

5.10 Adapted from Rokas, A. and Carroll, S. B. (2006). Bushes in the tree of life. *PLoS Biology* 4: e352.

6.2 Based on Sober, E. (1993 [1984]). *The Nature of Selection: Evolutionary Theory in Philosophical Focus*. Chicago, IL: University of Chicago Press, p. 99.

6.6 Adapted from Mallet, J. (2008). Hybridization, ecological races, and the nature of species: empirical evidence for the ease of speciation. *Philosophical Transactions of the Royal Society of London, B Biological Sciences* 363: 2971–2986; Mallet, J.,

Meyer, A., Nosil, P., and Feder, J. L. (2009). Space, sympatry and speciation. *Journal of Evolutionary Biology* 22: 2332–2341.

6.8 Adapted from Ptacek, M. B. and Hankinson, S. J. (2009). The pattern and process of speciation. In M. Ruse and J. Travis (Eds.) *Evolution: The First Four Billion Years*. Cambridge, MA: Harvard University Press, pp. 177–207, p.178.

6.9 Adapted from Jablonski, D. (2007). Scale and hierarchy in macroevolution. *Palaeontology* 50: 87–109, p. 91.

Index